Neuropsychology of the Sense of Agency

Michela Balconi (Ed.)

Neuropsychology of the Sense of Agency

From Consciousness to Action

Alliant International University
Los Angeles Campus Library
1000 South Fremont Ave., Unit 5
Alhambra, CA 91803

 Springer

Editor
Michela Balconi
Department of Psychology
Catholic University of Milan
Milan, Italy

ISBN 978-88-470-1586-9 e-ISBN 978-88-470-1587-6

DOI 10.1007/978-88-470-1587-6

Springer Milan Dordrecht Heidelberg London New York

Library of Congress Control Number: 2010925396

© Springer Verlag Italia 2010

This work is subject to copyright. All rights are reserved, whether the whole or part of the material is concerned, specifically the rights of translation, reprinting, reuse of illustrations, recitation, broadcasting, reproduction on microfilm or in any other way, and storage in data banks. Duplication of this publication or parts thereof is permitted only under the provisions of the Italian Copyright Law in its current version, and permission for use must always be obtained from Springer. Violations are liable to prosecution under the Italian Copyright Law.

The use of general descriptive names, registered names, trademarks, etc. in this publication does not imply, even in the absence of a specific statement, that such names are exempt from the relevant protective laws and regulations and therefore free for general use.

Product liability: The publishers cannot guarantee the accuracy of any information about dosage and application contained in this book. In every individual case the user must check such information by consulting the relevant literature.

9 8 7 6 5 4 3 2 1

Cover design: Ikona S.r.l., Milan, Italy

Typesetting: Graphostudio, Milan, Italy
Printing and binding: Arti Grafiche Nidasio, Assago (MI), Italy
Printed in Italy

Springer-Verlag Italia S.r.l. – Via Decembrio 28 – I-20137 Milan
Springer is a part of Springer Science+Business Media (www.springer.com)

Preface

*Not nothing
without you
but not the same*
Erich Fried (1979)

How do I know that I am the person who is moving? The neuroscience of action has identified specific cognitive processes that allow the organism to refer the cause or origin of an action to its agent. This sense of agency has been defined as the sense that I am the one who is causing or generating an action or a certain thought in my stream of consciousness. As such, one can distinguish actions that are self-generated from those generated by others, giving rise to the experience of a self-other distinction in the domain of action.

A tentative list of the features distinguishing the concept of agency includes awareness of a goal, of an intention to act, and of initiation of action; awareness of movements; a sense of activity, of mental effort, and of control; and the concept of authorship. However, it remains unclear how these various aspects of action and agency are related, to what extent they are dissociable, and whether some are more basic than others. Their sources remain to be specified and their relationship to action specification and action control mechanism is as yet unknown.

Certain cues must be considered as contributing to the awareness of action or its disruption. These include efferent or central motor signals, reafferent feedback signals from proprioception, vision, action intentions or prior action-relevant thoughts, primary knowledge, and signals from the environment. Of these, the experience of intentionality, of purposiveness, and of mental causation is of particular interest. Intention directly contributes to the sense of agency and in this volume is extensively discussed, e.g., with respect to the control of action and the afferent information coming from peripheral areas of our body.

Moreover, recent conceptual developments have provided new perspectives on the sense of agency, separating an implicit level of *feeling of agency* from an explicit level of *judgment of agency*. The first is thought to be characterized by lower-level, pre-reflective, sensorimotor processes, and the second by higher-order, reflective, or belief-like processes. In this view, sensorimotor processes contributing to the feeling level may run outside of consciousness but may be available to awareness. This is supported by empirical evidence that, for example, minor violations of intended actions or action consequences (i.e., brief temporal delays in sensory feedback) do not necessarily enter awareness, while neural signatures of such violations can be observed. Experimental operationalization of the sense of agency should consider these different levels of agency to systematically explore the multiple indicators of agency and their possible interplay. However, empirical investigations often focus on judgments or attributions of agency, involving subjective reports and plagued by errors through misidentification. By contrast, multivariate approaches that include implicit measures (kinematics, eye movements, motor potentials, brain activity, etc.) may better access the feeling level of agency.

This book offers an integrated perspective by considering the psychological, cognitive, neuropsychological, and clinical implications of agency. It consists of three main sections. The first, *Cognition, Consciousness, and Agency*, introduces the topic of agency, highlighting the main critical points of agency investigations. Specifically, the theoretical and empirical implications of the sense of agency for consciousness, self-consciousness, and action are considered in Chapter 1, which seeks a causal explanation of action and analyzes potential mechanisms underlying the conscious control of action, as implicated in normal individuals and in pathological cases. We also examine the role that intentions have for agency representation with respect to initiation, control, and action execution. Another point of interest is the "illusion of agency," which provides a critical perspective on the concept of free choice and the overt representation of self for action. Chapter 2 explores the sense of agency from a non-mentalistic view, assuming that agency is at least partly grounded in the perception of various kinds of affordances. Thus, there is a common perceptual component in awareness that we are doing something. This component is crucial to understanding our sense of *joint* agency, when we cooperate with other agents in order to achieve a shared goal.

The investigation of the sense of agency is an increasingly prominent field of research in psychology and in cognitive neurosciences, as is underlined in the second section of the book; *Brain, Agency, and Self-agency: Neuropsychological Contributions to the Development of the Sense of Agency*. Chapter 3 introduces this section by exploring recent research developments, which have approached the study of the sense of agency from the two levels mentioned above, i.e., the implicit feeling of agency and the explicit judgment of agency. Chapter 4 reviews the current neuropsychological and neuroimaging data, which together have suggested several different brain regions as key candidates in the functional anatomy of agency, thereby distinguishing the different aspects of the agency experience. The subjective significance of the sense of agency for consciousness is the focus of Chapter 5, which, by reporting empirical results, explores the relationship between fluctuations in daily experience and agency. Several studies have shown that agency includes an awareness of body movements, goals, intention in action, sense of activity, mental effort, and the control of action execution. These dimensions are also constituents of subjective experience, which results from the interplay between emotion, motivation, and cognition in response to internal and environmental events. The role of interaction for agency construct is explored in Chapter 6, beginning with a few philosophical remarks on agency and intentionality and then going on to examine the significance of joined intentions and joint action for the construction of inter-agency. Some of the most recent suggestions about joint agency are discussed, as is the relationship between agency and sociality as well as the implications of agency for social behavior and its function in human interactions.

The sense of agency may be disrupted in situations in which action feedback is unexpected or erroneous, as is critically explored in the book's third section, *Clinical Aspects Associated with Disruption of the Sense of Agency*. Chapter 7 proposes a synthetic view of the different deficits in agency, taking into account perceptive, attentive, and psychiatric domains. Pathologies such as blindsight and numbsense, visual and somatosensory neglect syndromes, and psychiatric disturbances (such as schizophrenia and autistic syndrome, as well as obsessive-compulsive disorders) include a consistent impairment in the sense of agency, from both the feeling and the judgment perspectives. A pilot electroencephalographic study on the perception of the anomalous feedback of action, in which agency is disrupted, is described. The results of that experiment have highlighted the role of personal differences, based on the individual's attitude toward action and on external cues, in the subjective response to mismatching conditions involving action and action feedback. Chapter 8 discusses the delusion of control in schizophrenia and in acute psychotic states, in which internal predictions about the sensory consequences of one's actions are imprecise. The sense of agency in psychotic patients is at constant risk of being misled by ad hoc events, invading beliefs, and confusing emotions and evaluations. Such patients might therefore be taught to rely more on alternative cues relating to self-action, such as vision, auditory input, prior expectations, and post-hoc thoughts, and to ignore the usually robust and reliable sources of internal action information. Taking into account the pathological and dysfunctional evolution of the sense of agency, Chapter 9 analyzes the mechanisms underlying the feeling of agency and the judgment of self-causation in obsessive-compulsive disorders. Finally, Chapter 10 considers the contribution of body representation in agency and self-awareness. The distinction between body ownership and agency is discussed, pointing out the significance of propioceptive feedback for self-representation in action. The chapter concludes by examining the disruption of the sense of agency such as occurs in classical syndromes, for example, "alien hand" and "anarchic hand" phenomena.

I would like to extend special thanks to Adriana Bortolotti, who contributed to the careful work of editing. The volume was partially funded by the Catholic University of Milan (D3.1. 2008).

Milan, June 2010 **Michela Balconi**

Contents

Section I	Cognition, Consciousness and Agency	1

1 The Sense of Agency in Psychology and Neuropsychology 3
Michela Balconi

    1.1   To Be an Agent: What Is the Sense of Agency? 3
    1.2   Action and Awareness of Action 4
    1.2.1   Does Awareness of Action Differ from the Sense of Agency? 5
    1.3   The Key Determinant Mechanisms for the Sense of Agency 6
    1.4   A Critical Approach to Intentions and Intentional
            Binding Phenomenon 9
    1.4.1   Awareness, Consciousness, and Agency: Unconscious
            Perception and Unconscious Intentions 11
    1.4.2   Self-consciousness and the Illusion of Agency 12
    1.4.3   Consciousness of Self and Consciousness of the Goal 14
    1.5   The Sense of Initiation 15
    1.5.1   The Limited Sense of Initiation: Libet's Contribution 15
    1.6   The Sense of Control 17
    1.7   The Sense of Agency for Self and for Others: The "Perceptual"
            Basis of Empathy 18
    References 20

2 Affordances and the Sense of Joint Agency 23
Jérôme Dokic

    2.1   Introduction 23
    2.2   Social Perception and Mind-reading 24
    2.3   The Concept of Affordances 28
    2.4   Instrumental *vs* Deontic Affordances 30
    2.5   Canonical Neurons as Reflecting Instrumental Affordances 31

2.6	Egocentric *vs* Allocentric Perception of Affordances	32
2.7	Mirror Neurons and Action-dependent Affordances	34
2.8	Interpersonal Affordances	36
2.9	Two Models of Joint Action	38
2.10	Conclusions	40
References		41

Section II Brain, Agency and Self-agency: Neuropsychological Contributions to the Development of the Sense of Agency 45

3 The Neuropsychology of Senses of Agency: Theoretical and Empirical Contributions ... 47
Michela Balconi

3.1	Different Types of the Sense of Agency	47
3.2	Feeling and Judgment in the Sense of Agency	49
3.3	Empirical Paradigms of the Judgment of Agency	51
3.3.1	The Awareness of Action: The Contribution of Event-related Potentials	51
3.3.2	Time Perception and the Sense of Agency	52
3.3.3	Visual Feedback and Awareness of Action	53
3.3.4	Somatosensory Information for Agency	55
3.3.5	Sense Integration	57
3.3.6	Experimental Paradigms for the Feeling of Agency	57
3.4	Minimal Self and Narrative Self	61
3.4.1	Minimal Self: Self-agency as "I"	61
3.4.2	Self Ascription	63
3.4.3	Narrative Self: The Sense of Continuity	63
References		66

4 Functional Anatomy of the Sense of Agency: Past Evidence and Future Directions ... 69
Nicole David

4.1	Introduction	69
4.2	A Functional Anatomy of the Sense of Agency: Past Evidence	70
4.2.1	Posterior Parietal Cortex and Inferior Parietal Lobule	72
4.2.2	The Cerebellum	73
4.2.3	The Posterior Superior Temporal Sulcus	74
4.2.4	The Insula	74
4.2.5	The Supplementary Motor Area	75
4.2.6	The Prefrontal Cortex	75
4.3	Future Directions	76
4.4	Conclusions	77
References		77

5	The Monitoring of Experience and Agency in Daily Life: A Study with Italian Adolescents 81	
Marta Bassi, Raffaela D.G. Sartori, Antonella Delle Fave		

	5.1	Agency and Its Role in Human Behavior and Experience 81
	5.2	Agency and Experience 84
	5.2.1	Defining and Measuring Experience 85
	5.2.2	Agency in Daily Life: A Crucial Component of Optimal Experience 87
	5.3	Empirical Evidence: A Study with Italian Adolescents 89
	5.3.1	Aims and Methods 91
	5.3.2	Results 91
	5.4	Agency and Daily Experience: A Promising Research Domain 96
	References 100	

6	Agency and Inter-agency, Action and Joint Action: Theoretical and Neuropsychological Evidence 107	
Davide Crivelli, Michela Balconi		

	6.1	Introduction 107
	6.2	An Introduction to Agency 108
	6.3	The Beginning: Intentions and Collective Intentions 109
	6.3.1	From I to We 109
	6.3.2	We in Action 110
	6.4	Doing Things Together: Joint Action and the Sense of Agency 111
	6.5	Over the Self-other Differentiation: Circular Interactions and Joint Agency 113
	6.5.1	The Intersubjective Origins of Joint Agency: A Developmental Perspective 115
	6.6	Inter-acting Selves, Social Agency, and Neural Correlates 116
	6.6.1	The Original Distinction of Our-selves and Other-selves 117
	6.6.2	Self-other Differentiation, Agency and Sociality: Hypotheses and Neuropsychological Evidence 118
	6.7	Conclusions 119
	References 120	

Section III	Clinical Aspects Associated with Disruption of the Sense of Agency 123

7	Disruption of the Sense of Agency: From Perception to Self-knowledge ... 125
Michela Balconi	

	7.1	Introduction 125
	7.2	Disruption of Agency in the Perceptual Field and in Proprioception 125

	7.2.1	Agency and Body: Predictivity Function of the Body for Self-representation	126
	7.2.2	Perceptual Illusions of Body	127
	7.2.3	Blindsight and Numbsense	128
	7.2.4	A Tentative Conclusion Regarding Perceptual Level Impairment	129
	7.3	Attentive Deficits and the Sense of Agency	129
	7.3.1	Visual Neglect Syndrome	129
	7.3.2	Somatosensory Neglect	130
	7.4	The Fallibility of Self-attribution of Agency in Neuropsychiatry	131
	7.4.1	Frontotemporal Dementia and the Delusion of Control in Frontal Deficits	132
	7.4.2	Agency and Schizophrenia	132
	7.4.3	Concluding Remarks on Schizophrenia	134
	7.4.4	Autism: Mentalizing *vs* Agency Disruption	135
	7.4.5	Dissociated States: Obsessive-compulsive Disorder	136
	7.4.6	Lines of Research on the Disruption of Agency: ERPs and Personality	137
	References		140

8 Disturbances of the Sense of Agency in Schizophrenia 145
Matthis Synofzik, Martin Voss

	8.1	Introduction	145
	8.2	The Comparator Model and Its Explanatory Limitations	146
	8.3	Feeling of Agency *vs* Judgement of Agency	148
	8.4	Optimal Cue Integration as the Basis of the Sense of Agency	150
	8.5	Altered Cue Integration as the Basis of Delusions of Influence	151
	8.5.1	Intentional Binding: Impaired Predictions and Excessive Linkage of External Sensory Events	151
	8.5.2	Perception of Hand Movements: Imprecise Predictions Prompting an Over-reliance on External Action Cues	152
	8.6	Conclusions	153
	References		154

9 Looking for Outcomes: The Experience of Control and Sense of Agency in Obsessive-compulsive Behaviors 157
Sanaâ Belayachi, Martial Van der Linden

	9.1	Introduction	157
	9.2	The Clinical Features and Phenomenology of OCD	158
	9.3	Sense of Agency in OCD: Empirical Data	160
	9.4	Summary and Discussion	164
	9.5	Conclusions	167
	References		168

10 Body and Self-awareness: Functional and Dysfunctional Mechanisms 173
Michela Balconi, Adriana Bortolotti

10.1	The Sense of Agency and the Sense of Ownership as Components of Self-consciousness	173
10.2	The Sense of Body Ownership *vs* the Sense of Agency	174
10.3	The Sense of My Body as Mine: A Threefold Perspective	175
10.4	A Spatial Hypothesis of Body Representation	178
10.5	Neural Substrates of the Sense of Ownership	180
10.6	Disruption of the Sense of Ownership: Conscious and Non-conscious Body Perception	182
10.6.1	The Rubber Hand Illusion: Evidence of Disownership Phenomena ..	184
10.6.2	Other Body Impairments: Neuropsychological Disorders	185
References		187

Subject Index 191

List of Contributors

Michela Balconi
Department of Psychology
Catholic University of Milan
Milan, Italy

Marta Bassi
Department of Preclinical Sciences
LITA Vialba
University of Milan
Milan, Italy

Sanaâ Belayachi
Department of Cognitive Science
University of Liège
Liège, Belgium

Adriana Bortolotti
Department of Psychology
Catholic University of Milan
Milan, Italy

Davide Crivelli
Department of Psychology
Catholic University of Milan
Milan, Italy

Nicole David
Department of Neurophysiology and
Pathophysiology
University Medical Center
Hamburg-Eppendorf
Hamburg, Germany

Antonella Delle Fave
Department of Preclinical Sciences
LITA Vialba
University of Milan
Milan, Italy

Jérôme Dokic
École des Hautes Études en Sciences
Sociales & Institut Jean-Nicod
Paris, France

Raffaela D.G. Sartori
Department of Preclinical Sciences
LITA Vialba
University of Milan
Milan, Italy

Matthis Synofzik
Department of Neurodegeneration
Center of Neurology & Hertie–Institute
for Clinical Brain Resarch
Tübingen, Germany

Martial Van der Linden
Department of Cognitive Science
University of Liège
Liège, Belgium

Martin Voss
Department of Psychiatry and
Psychotherapy
Charité University Hospital /
St. Hedwig Hospital
Berlin, Germany

Section I
Cognition, Consciousness and Agency

The Sense of Agency in Psychology and Neuropsychology

M. Balconi

1.1
To Be an Agent: What Is the Sense of Agency?

The sense of agency is an increasingly prominent field of research in psychology and the cognitive neurosciences. In this chapter, *awareness of action* is distinguished from the *sense of agency*, since they represent different elements of self-awareness and self-monitoring in action execution. Nevertheless, both contribute to causing or generating an action or a certain thought in the stream of consciousness. Here, we offer a causal explanation of action and address the mechanisms behind the conscious control of action, as they occur under normal and pathological conditions. Specifically, we consider the theoretical and empirical implications of the sense of agency for consciousness, self-consciousness, and action. The main question is how do I know that I am the person who is moving? Psychology and the neuroscience of action have shown the existence of specific cognitive processes allowing the organism to refer the cause or origin of an action to its *agent* [1]. This sense of agency has been defined as the sense that I am the one who is causing or generating an action or a certain thought in my stream of consciousness [2]. As such, one can distinguish actions that are self-generated from those that are generated by others, giving rise to the experience of a self–other distinction in the domain of action which, in turn, contributes to the subjective phenomenon of self-consciousness.

A tentative list of distinctions regarding the concept of agency includes awareness of a goal, of an intention to act, and of initiation of action, as well as awareness of movements, sense of activity, sense of mental effort, sense of control, and the concept of authorship. Yet, it remains unclear how these various aspects of the phenomenology of action and agency are related, to what extent they are dissociable, and

M. Balconi (✉)
Department of Psychology, Catholic University of Milan, Milan, Italy

whether some are more basic than others. Furthermore, their sources and how they relate to action specification and an action control mechanism remain to be specified. In the interactions with others, the experience of intentionality, of purposiveness, of mental causation must be considered.

From a neuropsychological point of view, there is evidence of different neural correlates for the sense of agency, which might reflect different agency indicators and/or sub-processes or levels of agency processing. The first group of brain areas involved in the sense of agency is located in the motor system and includes the ventral premotor cortex, the supplementary motor area (SMA) and pre-SMA, and the cerebellum. A second group consists of the dorsolateral prefrontal cortex (DLPFC), the posterior parietal cortex, the posterior segment of the superior temporal sulcus, and the insula (see also Chapter 4). It is probable that the first group constitutes a network of sensory-motor transformation and motor control, and the second a set of association cortices implicated in various cognitive functions. An example of the latter is the DLPFC, which could be relevant in various cognitive functions, such as behavior, in the temporal domain [3]. More generally, regions of the motor system may subserve executive functions, and heteromodal associative regions supervisory functions.

1.2
Action and Awareness of Action

It is widely assumed that the processes through which the component elements of the phenomenology of action are generated and those involved in the awareness and control of action are strongly interconnected. But what is an action? For Marcel [4], it is not only a bodily motion or a simple re-action to an external or internal stimulus; rather, it has a goal, end-point, or effect. We can distinguish the notion of intention, in which case an action is a realization, from the notions of intentional directedness and content. Quite apart from being the realization of an intention, actions are defined as such by their directness, as having definite end-points. Thus, an action is not a mere bodily movement but consists of two parts, the *movement* and the *intention-in-action* that causes that movement [5]. Moreover, an action has some degree of *voluntariness*. For this reason, it is necessary to distinguish an action from a habit and from something that is caused in the mechanical sense. A movement or behavior will be seen as an action to the extent that it is "agentive," i.e., that it is self-generated and performed at one's will.

Another distinction is between *physical actions* and *mental actions*. In general, physical actions involve the production of causal effects in the external world through movements of the body of the agent, while mental actions, such as pretending or remembering a name, do not. Thus, the phenomenology of physical actions can be viewed in terms of a sense of oneself as a physical agent producing physical effects in the world via bodily interactions with it. In Jeannerod's componential view of action [6], bodily movements are merely the overt part of actions that also necessarily involve a covert, representational part. Thus, what distinguishes an action from merely being a bodily movement is the fact that the person is in some particular relationship to the movements of his body during the time in which he is performing them and that this relation is one of *guidance*. In addition, the agent of an action is aware of what is he is doing by virtue of controlling the action, rather than on the basis of observation or introspection [7]. The relevant notion of *control* is that of rational control, which can be described as a matter of practical reasoning leading to action.

We can also distinguish between *what* an action is and *how* it is performed, or the goal pursued and awareness of the means employed to attain this goal. Specifically, the phenomenology of action itself concerns what is being done. Actions have a goal and they involve an element of purposiveness. In other words, we are aware to some degree that we are engaged in purposive activity. As to the issue of how, beyond being aware of the goal of our action, we have an awareness of the specific manner by which the desired result is being achieved.

1.2.1
Does Awareness of Action Differ from the Sense of Agency?

In the above discussion, a constitutive link between the agent's awareness of an action and his sense of agency was implied. Nevertheless, empirical evidence suggests that although *awareness of action* and *sense of agency* normally go together, they sometimes diverge. We can therefore ask whether awareness of an action performed by one's self is sufficient to impart the sense that it is one's own action. As pointed out by Dennett (1991) [8], we are not authoritative or incorrigible as to our conscious experience. There are many examples in which people are unaware of their phenomenology or are unable to be aware of it. For example, the constraints of attention make it hard to be aware of all of one's phenomenology at a time. Or, one may be generally aware of something without knowing exactly what comprises the experience, as is the case for emotional experience. In other words, I may be aware of performing a certain action without being able to tell the exact form the content of that awareness takes.

From a clinical point of view, a vast amount of data suggest, for example, that in schizophrenia the sense of alien control derives from the fact that the subject is aware of the content of the action he is executing but denies the agent of this action (see also Chapter 8) [9]. A dissociation between awareness of action and sense of agency can also occur in non-pathological conditions. *Illusion of control*, in which we experience a sense of agency for actions someone else is doing, and *illusion of action* projections, in which we do not experience a sense of agency for something we are doing, can be observed in normal subjects [10] (for further discussion of these concepts, see Chapter 7).

Many experimental investigations into the sense of agency *vs* the awareness of action have manipulated the sensory, and particularly the visual, consequences of a subject's actions. In the classical paradigm, a subject is asked to draw a line on a paper and while doing so is able to either see the result of his own hand movements or, unbeknownst to him, that of an "alien hand" (for example, the experimenter's). The alien hand's movements are spatially deviated from the subject's own movements. Generally, under these conditions, the subject adjusts his own actual movement to the false visual feedback without being aware of the adjustment [11-13]. However, the sense of agency cannot be considered as being solely influenced by *visual re-afferences*. Only a few experiments have manipulated internal signals such as proprioceptive or motor signals. The problem in such studies lies in the fact that if subjects are instructed to explicitly evaluate self–other agency, internal signals, such as intentions, as much as external signals, such as visual re-afferences, may influence the subject's judgments. Therefore, the question remains: what are the main differences between the awareness of action and the sense of agency for self?

We cannot reduce the sense of agency to the *sense of ownership* of one's body despite the wide importance of their interactions [2] (see also Chapter 10). An action is not only perceived but is also initiated, controlled, or inhibited. Consequently, we have to take into account the dimension of the agent who is the *cause* of the action. In this perspective, I may not be the agent of all my bodily movements, as in passive movements when, for example, someone else raises my arm for me. It is only in a second sense that passive movements are mine because all I own in these cases is the moving body. Therefore, what must be added to the definition of intention refers to the neutral state (performed action) and *the ability to self-ascribe it* ("I am moving").

The sense of agency is intimately linked to the *sense of causality* and it results from the intentional binding of *intentions*, *actions*, and *sensory feedback*, which are "attracted" to each other, reinforcing the perception of their causal relations. Therefore, to understand actions, we need to analyze their causal antecedents, that is, what initiates the action's occurrence ("why am I writing") and what specifies the content of the action ("why am I doing it so"). Moreover, actions are not only preceded by an intention that is independent of the execution, they are also continuously represented in the *intention-in-action* until the end of the action. The sense of agency is not only the experience of an act of will that is distinct from bodily movements: it is the *experience* of the *continuous control of action execution*. For example, in anosognosic patients the sense of initiation is disrupted, while de-afferented patients suffer from a deficit of the sense of their own movements. Anosognosic patients do not try to initiate any action and do not send any efference copy that could be compared to sensory feedback and that would inform them that the intended movement has not been performed (for the concept of sensory feedback, see Par. 1.3).

1.3
The Key Determinant Mechanisms for the Sense of Agency

A large body of evidence suggests that the sense of agency, especially the judgement of agency (also see Chapter 3), strongly depends on the degree of congruence *vs* incongruence between predicted and actual sensory outcome [14, 15]. Congruence of the predicted with the actual outcome leads to attribution of the sense of agency to oneself, whereas incongruence indicates another agent as the cause of an action. Figure 1.1 diagrams the congruence and incongruence of predicted outcomes and their effects.

Some cues must be considered as contributing to the sense of agency or its disruption: efferent or central motor signals, re-afferent feedback signals from proprioception, vision, action intentions or prior-action relevant thoughts, primary knowledge, and cues from the environmental context. The motor control system seems to make use of internal models that mimic aspects of the agent and of the external world. Internal models have been proposed as being directly linked to the concept of control strategies [16]. The two main types of internal models suggested for human motor control are: (1) *forward models*, which mimic or represent the causal flow of a process in a system and use it to predict the next state of that system, and (2) *inverse models*, which compute the motor commands that have to be carried out to move a system from its current state to the desired one [17]. Common to these two models is the concept of *comparator*, which can be defined as comprising the mechanisms that compare two signals and use the result for the system's regulation (Fig. 1.2). Similarly, the prevailing explanation of the sense of agency of our own actions is the *central monitoring theory*, also called the *comparator model*, which postulates that the monitoring

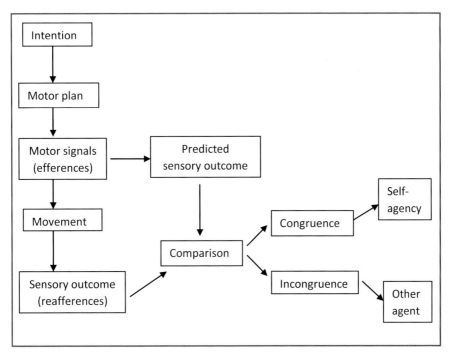

Fig. 1.1 Central monitoring theory or the comparator model. (Partially modified from David et al. [21])

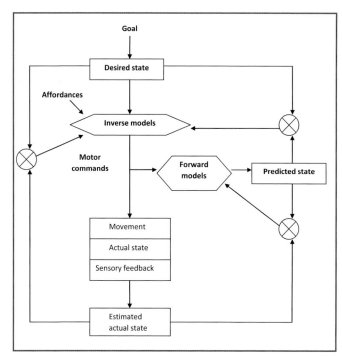

Fig. 1.2 An example of the comparator model

of central and peripheral signals arises as a consequence of the execution of an action. In this theory, (central) efferent signals at the origin of an action are matched with those resulting from its execution (the re-afferent signals), with the comparison providing cues about where and when the action originated [17-19].

However, the general validity of the comparator model was recently challenged. Synofzik, Vosgerau, and Newen [20], for example, found that the sense of agency can be determined by factors independent from any specific comparator output at the level of action control and action perception. Instead, awareness and attribution of agency were suggested to critically rely on *higher-order causal inferences* between thoughts and actions on the bases of belief states and intentional stances [10]. A clear example of this is that we often perform movements that we acknowledge were incongruent to our prediction. Yet we do not attribute their cause to an external origin, but still recognize ourselves as their agents. In this case, agency is inferred on the basis of *higher-order cognitive processing* that exploits environmental and contextual cues but is not evoked by an efference copy or any comparator output.

The *simulation model* invokes a functional role of the motor system and offers an alternative to the comparator model [21]. It proposes that, in understanding or predicting other people's behavior, we use our own experiences to simulate those of others [22]. Nonetheless, assuming shared representations for self and other cannot explain why we normally do not confuse our own actions with those of other people. Since the mirror neuron system does not provide an explicit representation of other agents, an additional mechanism must be assumed in which the representation of me or someone else as the agent is recognized.

Other constructs must be included in the determination of the sense of agency, an important one being the *intention* of action. Specifically, three principal mechanisms of intention intervene in defining the sense of agency for an actor: the *sense of intentionality* or intentional causation, based on the efferent binding of an action and its effect; the *sense of initiation*, which binds intention and movement onset; and *the sense of control* of actions and thoughts. All three are examined in detail in the following.

1.4
A Critical Approach to Intentions and Intentional Binding Phenomenon

While intentionality can be seen as coinciding with the sense that the agent is the cause of that action, i.e., a sense of intentional causation, at a more abstract level we can feel that our intention is the cause of our action, and at a lower level that our movement is causing some effect [23-25]. Thus, there is a match between prior intention and an observed action.

The experience of consciously willing our actions seems to arise primarily when we believe that our thoughts have caused our actions. We experience ourselves as agents of our actions when our minds provide us with previews of the actions that turn out to be accurate when we observe the action that ensues. Nevertheless, this experience may not be veridical, in that agents may have an experience of agency of an action they have not actually caused or, conversely, attribute their own actions to others. A significant example of the dissociation between the sense of agency and intentionality derives from the observation that matches other than just the match between a prior intention and an observed action are important as well. One such match is between an action and its consequences. Usually, the perceived time at which an action is initiated is experienced as being closer to the perceived time of the effect. In other terms, the action is shifted forwards in time towards the effect it produces, while the effect is shifted backwards in time towards the action that produced it (*intentional binding*). This observation may be used as an implicit measure of the sense of agency, serving as a predictive and inferential mechanism of action control. This mechanism depends critically on the presence of voluntary movement and requires an efferent signal; the experience of intentional causation appears to be constructed at the time of the action itself, as an immediate by-product of the motor control circuits that generate and control physical movement; accordingly, there must be a reliable relationship between actions and effects.

In several experiments Haggard and colleagues found support for the idea that voluntary, but not involuntary, movements and movement consequences are temporally bound together in conscious awareness [26, 27]. In those studies, subjects judged the perceived onset of voluntary movements as occurring later and the sensory consequences as occurring earlier than was actually the case. Once an intention-to-act has been formed, actions and action consequences are more likely to be attributed to oneself even if they were externally generated. This can be understood as a sense of intentional causation and as an intentional binding phenomenon between an individual (agent) and the observed action of others. It is manifested as the tendency of subjects to naturally perceive themselves as being causally effective.

The influence of action-relevant thoughts that increase feelings of self-efficacy over movements was investigated by Wegner and colleagues [28]. According to those authors, intentional binding is a link between intention and action and serves many functions. It may be important during motor learning, for example [29]: I can learn to correct an error if I can associate it with the corresponding intention. This approach may be useful in the construction of the agent-self, as it confers the ability to relate the content of one's intentions to actions and their environmental consequences. It is directly linked to the tendency of subjects to naturally perceive themselves as causally effective and proficient [30].

Intentional binding may be related to increased activation of the SMA or pre-SMA and insula. These areas have been associated with awareness and the execution of self-generated actions, with action preparation, and with the subject's own intention-to-act [31]. Evidence for the relevance of the supplementary motor cortex to the experience of intentional actions also comes from patients with neurological conditions: lesions in the SMA have been associated with the so-called anarchic hand syndrome, in which patients experience unintended actions of their own hand just as if the hand had an "independent will." The binding effect also occurs when we observe other people's actions. This was concluded from the results of experiments in which subjects had to estimate the onset time of pressing a button, which they executed themselves or which they observed being executed by someone else or by a mechanical device. The estimate of the machine actions was always different from those of self- or other-generated actions, whereas the latter two were indistinguishable. Subjects had a slightly delayed awareness of the onset of their own actions and of the experimenter's action, evidencing in both cases a binding effect.

These findings are inconsistent with the predictive account of intentional binding favored by Haggard [26], provided one assumes that the predictive mechanisms used for action control also operate when one observes someone else acting. If intentional binding is not linked to a particular person, it cannot be the basis of the sense of authorship for an action. Thus, intentional binding of action and effect would seem to be associated with the agent-neutral experience of intentional causation, rather than with the experience of authorship per se [32].

More generally, the sense of intentional causation cannot be the unique and primary factor of the sense of agency; instead, it is a necessary but not sufficient component in the generation of a sense of agency in so far as it can be present when one observes actions performed by other agents. For example, we often cannot remember what our prior intentions were and yet we do not disown the actions. It is not clear how this effect can effectively support the sense of "I" for action, since a binding effect and sense of intentional causation also occur when we observe other people's actions.

1.4.1
Awareness, Consciousness, and Agency: Unconscious Perception and Unconscious Intentions

Broad discussions of the relationship between intentions and agency, focusing on the role of intentional goal a subject has for action, are not found in the literature. We view intentions as the main factor establishing the link between self and action, although we do not consider intention and intention awareness as a prerequisite for the sense of agency. Conscious representation of our own actions is not a condition sine qua non for the agency's existence since also unconscious contexts may generate the sense of authorship and ownership that constitute subjective agency. This may explain the "illusion of agency," especially in those cases in which the sense of agency is inferred from the attribution of self as an agent who has intentionally produced that action, when the person did not really cause it.

The awareness of intentions cannot be necessary for awareness of action and agency: on many occasions, we cannot remember why we are doing something, thus neither denying its status as an action nor disowning it. Recent work suggests that the affordances of an object or situation are automatically detected even in the absence of any intention to act. These affordances automatically activate corresponding motor programs [33]. One of the main features of the motor system is its limited cognitive penetrability such that some global aspects of its operation appear to be consciously accessible and to be reflected in conscious motor imagery [34] whereas we are not aware of the precise details of the motor commands that are used to generate our actions. Moreover, the construct of "effort" may not be adequate to explain the preconditions of awareness of action and the sense of agency. The degree of effort derived from executing an action is not sufficient to support the awareness of being effective in carrying out an action. There are cases of virtually effortless actions of which we are not unaware.

Likewise, proprioceptive awareness cannot be necessary for an awareness of action. Some patients who are deprived of all proprioceptive experience and bodily sensation are not aware of their actions without vision of the disposition of their limbs and body, but they know that they have acted. Neither can proprioceptive awareness be sufficient for people to experience action: for example, reflex passive movements are proprioceptively experienced but they are not experienced as actions.

More generally, to what extent are we aware of our intentions, and to what extent is this awareness necessary to represent our sense of agency? These questions are a core point since they contribute to assessing the role of awareness of intention in our own awareness and in the ownership of action. In addition, an awareness of intention may contribute to having a long-term sense of agency.

Often it is assumed that intentions are by their nature conscious. Nevertheless, there are several ways in which we may be unaware of them. Firstly, we frequently lack explicit awareness of sub-goals in achieving a goal. This applies not only to how one is achieving the goal with respect to the manner of an action, but also to actions that are instrumental for the overall goals. Even when I perform a sub-goal first, as a discrete action, I may be unaware of intending it. However, in most cases one can become aware of such intentions. Secondly, in the course of temporally extended actions, we may forget our intention or the reason behind it. These cases often consist of the awareness of the sub-goal with temporary unawareness of the goal. A third example regards non-conscious long-term intentions whose presence is implied by the effect of their violation. These may be seen more as dispositional concerns than as intentions. Moreover, in immersed ongoing action, where we are not in a detached, self-reflective state, we have a general sense of acting intentionally but we not are aware of each intention and are often unaware of the specific content of each intention.

It is also important to emphasize the unawareness of action, rather than intentions. The problem is to determine the content of the conscious level, which is not merely a reading of the content of the automatic level of action (for this distinction, see [35]). The content of the conscious level does not include the complete set of details regarding what has actually been performed and how it has been performed. Introspectively, the agent seems to have access only to the general context of the action, i.e., its ultimate goal, its consequences, and the possible alternatives to it.

1.4.2
Self-consciousness and the Illusion of Agency

There are several kinds of self-consciousness. For example, the awareness of oneself may be primarily as a mental or as a physical entity. Self-consciousness may be long-term, persisting overtime while in others cases it is current and present-tense. There may be a detached awareness of oneself, or a more immediate or immersed one. Self-consciousness may also be the sense of self as a physical agent; that is, the sense of being an entity that exists in the physical world and has physical effects via one's physicality. But there is also a second sense, of a mental self that is a non-physical realization, such as the experience of one's intentions as one's own.

Is the concept of consciousness, whether physical or mental, necessary for the sense of agency? Some but not all of the processes of action production and agency depend on conscious experiences. In fact, generally, a basic form of self-consciousness (awareness or the attribution of "who" to the action) is not informed by conceptual thoughts or reflective processing [2]. Must we therefore conclude that agency is an illusion, unconsciously determined? In general, it is important for people to feel that they are captain of doing what they are doing. Subjects are profoundly interested in maintaining the fiction that they have conscious will, and this illusion seems to have positive effects concerning health [36]. Indeed, retrospective construction of the feeling of free choice occurs especially in those cases in which we are uncertain about the degree of deliberateness of an action.

Wegner [10] introduced two principles that may explain the experience of causing an effect: the *priority principle* states that "if you think about an entity just before some event happens to it, you tend to believe that your thought caused the event." The *consistency principle* states that "if your thoughts about an entity are consistent with what happens to it, you tend to believe that you caused what happened to it." In this view, action and conscious thoughts are represented as being produced in parallel, both being generated by the real driving force, which are unconscious neural events (Fig. 1.3).

The link between conscious thoughts and action represents a causal path that does not occur in reality but may be erroneously inferred, in the same way as any external causal relationship is inferred [36]. Retrospective reconstruction assumes a predictive mechanism of the phenomenal experience of intention. According to Wolpert and Ghahramani [37], a forward model makes predictions about the behavior of the motor system and its sensory consequences. These predictions are used to compare the actual outcome of a motor command with the desired outcome, enabling rapid error correction before sensory feedback is available. In line with that model, sensory attenuation has been shown to result from these kinds of predictive mechanisms. Similar processes are implicated when subjects are led to believe that they consciously intended actions or consequences of actions they did not produce themselves, and are based on the mechanism of back referral of an intention [36, 38]. Recent empirical studies demonstrated that subjects always indicate that they intentionally initiated an action while reaction time data strongly suggest that they in fact failed to stop the action or they misattributed their awareness of intention as a function of intentional involvement during action planning [36].

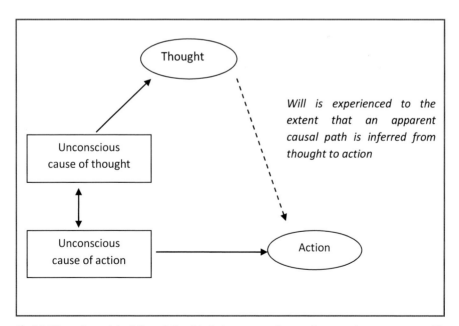

Fig. 1.3 Wegner's model of the relationship between conscious and unconscious processes with respect to the generation of voluntary actions

1.4.3
Consciousness of Self and Consciousness of the Goal

Some important distinctions should be made about the consciousness of *self in action* and the consciousness of *action per se*. The former is related to being conscious of oneself as causal self. The latter is related to what the action is about in terms of goals. To be aware of the goal one strives for is one way of being conscious of the action undertaken to reach that goal.

In addition, there may be a contrast between the overt and covert aspects of the goal. Whereas the detailed target of the movement remains outside consciousness, the overt goal of the actions, concerning the selection of objects, their use, their adequacy for the task under execution, etc., can be consciously represented. The covert aspects of the goal can, nonetheless, be consciously accessed. For example, in a series of experiments, subjects were instructed to indicate the moment at which they became aware of a change in the configuration of a target occurring during their movement. An analogous situation is when we are driving a car and have to change its trajectory because of a sudden obstacle in the road: we consciously see the obstacle after we have avoided it [6, 39]. In general, the awareness of a discordance between an action and its sensory consequences emerges when the magnitude of the discordance exceeds a certain amount.

The view of consciousness that arises when consciousness is related to action is a lengthy process that can take place only if adequate time constraints are fulfilled. Secondly, the type of consciousness that is linked to the experience of embodied self is discontinuous, operating on a moment-to-moment basis and bound to particular bodily events. The embodied sense of agency carries an implicit mode of action consciousness, in which consciousness becomes manifest only when required by the situation at hand. The information derived from this experience generally is short-lived and does not survive the bodily event for very long. By contrast, the sense of consciousness that we experience when we execute an action gives us a feeling of continuity, arising from the belief that our thoughts can have a causal influence on our behavior. Nevertheless, we generally ignore the cause of our actions but perceive ourselves as causal. The dissociation between the two levels of the self, the *embodied self* and the *narrative self* (see Chapter 3), has been considered as the origin of an illusion: the narrative self tends to build a cause-effect explanation, whereas the embodied self, by avoiding conscious introspection, reaches simpler conclusions about an action, its goal, and its agent by on-line monitoring of the degree of congruence between the central and peripheral signals generated by the action. Overall, the role of consciousness, both the short- and the long-term type, is to ensure the continuity of subjective experience across actions.

1.5
The Sense of Initiation

The sense of initiation comes from the link between intentions and movements, and it is reported by the subject between 80 and 200 ms before the movement actually occurs [40]. Specifically, the coexistence of *awareness of intention* and *awareness of movement onset* within a single narrow window of pre-motor processing suggests that the binding of these two representations is important: efferent binding may underline the sense of initiation for the action, such that the sense of initiation is not just the sense that we have started moving, but that we did so in accordance with our intentions.

Both intention judgments corresponding to the awareness of an intention to move and movement judgment corresponding to the awareness of movement onset precede actual movement and were found to co-vary with the *lateralized readiness potential* (LRP) (see Par. 1.5.1) effect of the event-related potential (ERP) [41]. This suggests that both awareness of intention and awareness of movements are associated with pre-motor processes rather than with the motor processes themselves. From a neuropsychological point of view, Sirigu and colleagues [42] showed that patients with parietal damage could report when they started moving but not when they first became aware of their intention to move, while cerebellar patients behaved like normal subjects. Both the parietal cortex and the cerebellum are thought to be involved in the predictive control of action: the cerebellum makes rapid predictions about the sensory consequences of self-generated movements, while the parietal lobe is implicated in high-level prediction and the more cognitive aspects of action.

The initiation of movements can therefore be said to have specific neural correlates, such as the dorsolateral prefrontal cortex (DLPFC), pre-SMA, SMA, basal ganglia, and primary sensory cortex. There is a great deal of parallel processing in this region, and the exact temporal order of activation of these areas when a movement starts is controversial, but it is a reasonable assumption that activity flows from the prefrontal region (DLPFC) in a generally caudal direction to finish in the primary motor area. In between, it reverberates around at least five separate cortico-basal ganglia loops, all of which are probably active in parallel. The output from this region is via the direct connections that exist between most of these areas and the spinal cord, which provides the final common pathway to the muscles.

However, it should be noted that although the sense of initiation may be a crucial component of the sense of agency for an action, it does not offer the guarantee that the whole action will be owned by the subject who performs the action. In some cases, we feel that we initiated an action but do not control its course.

1.5.1
The Limited Sense of Initiation: Libet's Contribution

Libet and colleagues [41] proposed that conscious awareness comes before the actual movement, but after the start of the brain activity leading up to it. They recorded the ERPs that preceded a voluntary finger movement and compared their time of onset with the time at which the subject reported becoming conscious that he was about to make each movement. Thus, it was possible that consciousness did not cause the movement in this particular case. Based on their findings, the authors suggested that, since consciousness arises slightly in advance of the movement, it is still capable of exerting a veto in the fraction of a second before the movement is executed. These results pointed out that some self-initiated and voluntary movements are triggered not by consciousness, but by unconscious actions of the brain. The subjects in those experiments reported experiencing conscious awareness of an "urge, intention or decision" to move approximately 200 ms before the movement actually took place. This event places the appearance of their conscious awareness that they were about to move exactly in the middle of what Libet called the type II readiness potential (RP) and other investigators the lateralized readiness potential (LRP). Although type II RPs can occur on their own if the subject has taken care not to pre-plan a particular movement, they can also be seen as the latter half of type I RPs, which begin a second or more before movement. The first segment of a type I RP, from the start of the waveform till about 500 ms before the movement, is probably underpinned by bilateral activity in the midline SMA. Type II RPs begins about 500 ms before the movement. Most EEG registrations put neural activity taking place between -500 and the movement as occurring mainly in the contralateral primary sensorimotor area (MI), with some residual activity still going on in the SMA. However, SMA activity was recorded in the interval between -300 and -100 ms and premotor cortex activity from -100 ms until onset of the movement, while MI activity was seen only from the onset of the movement until 100 ms afterwards. This would suggest that the neural generators of the type II RP lie in the SMA rather than in MI (Fig. 1.4).

In parallel, Wegner [10] proposed that the experience of having caused an action is not different from any other experience of cause-effect: in other words, we maintain that something else causes a certain effect if what we think of as the causal event occurs just before what we think of as the effect (the *priority principle*) and is the only apparent cause of the effect (the e*xclusivity effect*) (see also Par. 1.4.2). The most famous of Wegner's experiments illustrated the functioning of these effects. Specifically, subjects were asked to move a cursor around a computer screen and every 30 s to stop the cursor over some object presented on the screen. In some cases, the effective movements were manipulated not only by the subject but also by the experimenter. Subjects were asked to give an "intentionality" rating to the movements (completely sure to have caused or to have not caused the movement). In both cases (subject or experimenter assigned with causing the movement), performance was quite poor. When the subject truly caused all of the stops, the average intentionality rating was only 56%. When the stops were actually forced by the experimenter, the intentionality rating was 52%.

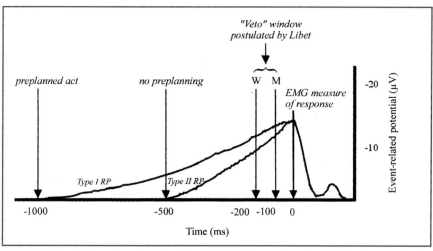

Fig. 1.4 Display of the main motor potentials (ERPs). *M*, motor intention; *EMG*, electromyogram; *W*, will intention

1.6
The Sense of Control

The sense of control can be represented as being made up of more basic, partly dissociable experiences. It may refer to the extent to which one *feels in control* of an action, such that the agent may feel that everything happens exactly as expected and he is in full control of his action. But it also may refer to sense that one has to *exert control*, to generate and maintain an appropriate action program. Generally, feeling in control is perceived as effortless, whereas the exertion of control is perceived as effortful and the effort increases in the case of the unexpected situation. The type of control one has to exert depends on the nature of the perturbing factors: external or internal, physical or not, anticipated or not. Generally, the more one feels that one is in control, the less one feels one has to exert control and vice versa.

Three more basic experiences may compound the sense of control: the sense of *motor control,* the sense of *situational control*, and the sense of *rational control.* In all three, the degree to which one feels in control depends on the results of a comparison between predicted and actual states; the better the match, the stronger the sense of control. One important difference between motor control on the one hand and situational and rational control on the other is that when one does not feel in full motor control one is simply aware that something is wrong, whereas when one does not feel in full situational or rational control, then one can be aware not just that something is wrong but of what is wrong.

The sense of control and feeling of mental effort are dissociable. Naccache and colleagues [43] observed that some patients are unable to be conscious of mental effort and the lack coincides with a lack of the bodily mediated physiological responses that index mental effort in healthy subjects. However, the sense of control for an action depends on comparisons between predicted states and actual states at various levels of action specification. We typically experience a feeling of effortless control when we achieve a perfect match between action and goal, i.e., without having to go through corrections or adjustments. Our sense of agency is heightened since the performed action fully conforms to our intention. In such actions, we meet no resistance and do not experience the kind of contrast between what we want and what the world will allow, which would sharpen our sense of self. In actions in which we meet resistance and have to overcome perturbations, the actual consequences of our actions do not match our predictions perfectly and we are left with the feeling that what we did was not exactly what we wanted to do. Nevertheless, at the same time our awareness of the efforts we have to make to try and keep the action on track heightens our sense that we are engaged in action.

1.7
The Sense of Agency for Self and for Others: The "Perceptual" Basis of Empathy

Some processes are related to the sense of agency, such as *imitation* and *perspective taking*, and to more basic, domain-general processes, such as *executive functions* and attention. Imitation and perspective taking imply a distinction between oneself and others. Both a first-person perspective and a sense of self agency have been proposed as key constituents of self-consciousness [2, 44]. Moreover, viewpoint-specific spatial cues have been discussed as indicators for the sense of agency; that is, knowing where the body is and what tools are available help to determine what the person could have authored. Other actions are generally associated with allocentric as opposed to egocentric representations.

Another experimental context that illustrates the dissociation between self-generated and other-generated action was based on auditory stimulation. Previous evidence in human subjects suggested that auditory stimuli are processed differently depending on whether they are a consequence of self-generated action. Shafer and Marcus [45], for example, showed that the cortical potentials evoked by self-produced tones have significantly smaller amplitudes and faster component latencies than those produced by a separate machine. More recently, Blakemore, Rees, and Frith [46] found that both predictability and the self-generated action have an effect on ERP modulation. The study demonstrated that different cortical areas are implicated in predictability *vs* self-generation of action, Specifically, the effect of hearing an auditory stimulus depends not only on its predictability, but also on whether the stimulus is produced by self-generated movement. The two effects are not simply additive since there is a modulatory effect of motor activity on stimulus predictability.

In line with previous considerations, the transparency property of subjective experience states that it is relatively *transparent* or obvious to the subject that the states disclosed by bodily experience are or are not of the same type as the states he can observe through external perception [47]. At one level of conscious experience, the sensations and actions of others are presented to the individual in much the same way as he is aware of his own sensations and actions. The transparency effect explains how empathy, imitation, and coordination are possible, in so far as these competencies depend on the perceptual ability to compare one's sensations and actions with those of others. A second relevant aspect regards the type of knowledge underlying both our own behavior and the behavior of others; that is, the fact that there is an asymmetry between the first-person and the third-person perspective: I may be aware of the sensations and actions of others from the outside, by observing their behavior, whereas I do not need observation in my own case (for the concept of I-thoughts, see Chap, 3) since I know from the "inside" that, for example, I am in pain.

Nevertheless, an important question is *how* the bodily experience of sensation and action may be transparent with respect to external perception. Recent models have tried to answer to this question by not assuming different modes of perception for internal *vs* external experience, or that the two types of experience involve different ways of experiencing the world. Rather, what grounds the sense of ownership is a constitutive relation between bodily experience and its intentional object, which makes the experience implicitly *reflective* [48]. Shifting the object of analysis from self-perception to perception of the other, recent models have proposed a substantial analogy between the two levels, stating that *social cognition* (or cognition of others from a perceptual or an action point of view) may be considered as a type of *extended field* of the subjective experience.

A role for direct perception in social cognition was defended by Gallagher [49]. The theory invoking direct perception is quite different from the standard theories of social cognition elaborated in psychology and cognitive science, the *theory theory* (TT) and *simulation theory* (ST). Both posit something more than a perceptual element, i.e., "mind-reading" or "mentalizing," as being necessary for our ability to understand others. By contrast, certain phenomenological approaches, such as described by the direct perception model, depend heavily on the concept of perception and the idea that we have a direct perceptual grasp of the other person's intentions, feelings, thoughts, etc. Both the TT and ST start with an understanding of perception as a third-person process, an observation of the other person, but each theory adds to perception certain cognitive elements that allow us to understand the other that we observe. Specifically, the TT contends that the way in which we understand other people depends on a practice of mentalizing, in which we employ a theory about how mental states inform the behavior of others. The ST claims that we have no need for theories like this because we have a model, in the form of our own mind, that we can use to simulate the other person's mental states. We begin by observing the other person's behavior in specific environments and, by simulation, we go on to model their beliefs and desires as if we ourselves were in their situation.

The concept of inter-subjective perception involves a relatively sophisticated process. It has been shown that young infants are visually attracted to movement, and in specific ways in the case of biological movements. For example, infants vocalize and gesture in a manner that seems tuned to the vocalizations and gestures of the other person [50]. Without the intervention of theory or simulation, and in a non-mentalizing way, they are able to see bodily movement as being expressive of emotion and as goal-directed intentional movement, in addition to being able to perceive other people as agents. This does not require advanced cognitive abilities, inference, or simulation skills but is a perceptual capacity that is fast, automatic, and highly stimulus-driven [51]. According to this general model, the mirror resonance mechanism (see [52]) may be thought as part of the structure of the personal process when it is a perception of another person's actions. In other words, it is hypothesized that mirror activation is not the initiation of simulation but it subtends a direct inter-subjective perception of what the other is doing [49].

References

1. Georgieff N, Jeannerod M (1998) Beyond consciousness of external reality: a "Who" system for consciousness of action and self-consciousness. Conscious Cogn 7:465-477
2. Gallagher S (2000) Phylosophical conceptions of the self: implications for cognitive science. Trends Cogn Sci 4:14-21
3. Vogeley K, Kupke C (2007) Disturbances of time consciousness from a phenomenological and a neuroscientific perspective. Schizophrenia Bull 33:157-165
4. Marcel AJ (2003) The sense of agency: awareness and ownership of action. In: Roessler J, Eilan N (eds) Agency and self-awareness: issues in philosophy and psychology. Oxford University Press, Oxford, pp 48-93
5. Searle J (1983) Intentionality. Cambridge University Press, United Kingdom Cambridge
6. Jeannerod M (2006) Motor cognition. Oxford University Press, United Kingdom Oxford
7. Balconi M, Santucci E (2008) Neuropsychological processes of motor imagery compared with motor scripts and action verbs. ERPs applied to motor representation. Proceedings of the First Meeting of the Federation of the European Societies of Neuropsychology (ESN), United Kingdom Edinburgh, pp 139
8. Dennett DC (1991) Consciousness Explained. Little, Brown and Company, Boston
9. Jeannerod M (2009) The sense of agency and its disturbances in schizophrenia: a reappraisal. Exp Brain Res 192:527-532
10. Wegner DM (2002) The illusion of conscious will. MIT Press, Cambridge, MA
11. Daprati E, Franck N, Georgieff N et al (1997) Looking for the agent: an investigation into consciousness of action and self-consciousness in schizophrenic patients. Cognition 65:71-86
12. Nielsen T (1963) Volition: a new experimental approach. Scan J Psychol 4:225-230
13. Slachevsky A, Pillon B, Fourneret P et al (2001) Preserved adjustment but impaired awareness in a sensory-motor conflict following prefrontal lesion. J Cognitive Neurosci 13:332-340
14. Fourneret P, Jeannerod M (1998) Limited conscious monitoring of motor performance in normal subjects. Neuropsychologia 36:1133-1140
15. Vosgerau G, Newen A (2007) Thoughts, motor actions, and the self. Mind Lang 22:22-43
16. Frith CD, Blakemore SJ, Wolpert DM (2000) Abnormalities in the awareness and control of action. Philos T R Soc Lon B 355:1771-1788
17. Blakemore SJ, Wolpert DM, Frith CD (2001) The cerebellum is involved in predicting the sensory consequences of action. Neuroreport 12:1879-1884
18. Blakemore SJ, Wolpert DM, Frith CD (1998) Central cancellation of self-produced tickle sensation. Nat Neurosci 1:635-640
19. Frith CD (1992) The cognitive neuropsychology of schizophrenia. Lawrence Erlbaum Associates, Hove

20. Synofzik M, Vosgerau G, Newen A (2008) Beyond the comparator model: a multifactorial two-step account of agency. Conscious Cogn 17:219-239
21. David N, Newen A, Vogeley K (2008) The "sense of agency" and its underlying cognitive and neural mechanism. Conscious Cogn 17:523-534
22. Goldman AI (1989) Interpretation psychologized. Mind Lang 4:161-185
23. Humphrey N (1992) A history of the mind. Simon & Schuster, New York
24. Wegner DM (2005) Who is the controller of the controlled processes? In: Hassin R, Uleman JS, Bargh JA (eds) The new unconscious. Oxford University Press, New York, pp 9-36
25. Pacherie M (2008) The phenomenology of action: a conceptual framework. Cognition 107:179-217
26. Haggard P, Clark S (2003) Intentional action: conscious experience and neural prediction. Conscious Cogn 12:695-707
27. Haggard P, Clark S, Kalogeras J (2002) Voluntary action and conscious awareness. Nat Neurosci 5:382-385
28. Wegner DM, Wheatley T (1999) Apparent mental causation. Sources of the experience of will. Am Psychol 54:480-492
29. Haggard P (2003) Conscious awareness of intention and of action. In: Roessler J, Elian N (eds) Agency and self-awareness: issues in philosophy and psychology. Oxford University Press, Oxford, pp 111-127
30. Wegner DM, Sparrow B (2004) Authorship processing. In: Gazzaniga MS (ed) The new cognitive neuroscience, 3rd edition. MIT Press, Cambridge, MA, pp 1201-1209
31. Cunnington R, Windischberger C, Robinson S, Moser E (2006) The selection of intended actions and the observation of other' actions: a time-resolved fMRI study. Neuroimage 29:1294-1302
32. Frith CD (2005) The self in action: lessons from delusions of control. Conscious Cogn 14:752-770
33. Grèzes J, Decety J (2002) Does visual perception afford action? Evidence from a neuroimaging study. Neuropsychologia 40:212-222
34. Jeannerod M (1994) The hand and the object: the role of posterior parietal cortex in forming motor representations. Can J Physiol Pharmacol 72:535-541
35. Jeannerod M (2003b) Consciousness of actionand self-consciousness: a cognitive neuroscience approach. In: Roessler J, Elian N (eds) Agency and self-awareness: Issues in philosophy and psychology. Oxford University Press, Oxford, pp 128-149
36. Kühn S, Brass M (2009) Retrospective construction of the judgement of free choice. Conscious Cogn 18:12-21
37. Wolpert DM, Ghahramani Z (2000) Computational principles of movement neuroscience. Nat Neurosci 3:1212-1217
38. Pronin E, Wegner DM, McCarthy K, Rodriguez S (2006) Everyday magical powers: the role of apparent mental causation in the overestimation of personal influence. J Pers Soc Psychol 91:218-231
39. Castiello U, Paulignan Y, Jeannerod M (1991) Temporal dissociation of motor responses and subjective awareness: a study in normal subjects. Brain 114:2639-2655
40. Libet B (1985) Unconscious cerebral initiative and the role of conscious will in voluntary action. Behav Brain Sci 8:529-566
41. Libet B, Gleason CA, Wright EW, Pearl DK (1983) Time of conscious intention to act in relation to onset of cerebral activities (readiness potential): the unconscious initiation of a freely voluntary act. Brain 106:623-642
42. Sirigu A, Daprati E, Ciancia S et al (2004) Altered awareness of voluntary action after damage to the parietal cortex. Nat Neurosci 7:80-84
43. Naccache L, Dehaene S, Habert MO et al (2005) Effortless control: executive attention and conscious feeling of mental effort are dissociable. Neuropsychologia 43:1318-1328

44. Metzinger T (2000) The subjectivity of subjective experience: a representationalist analysis of the first-person perspective. In: Metzinger V (ed) Neural correlates of consciousness. MIT Press, Cambridge, MA, pp 285-306
45. Shafer WP, Marcus MM (1973) Self-stimulation alters human sensory brain responses. Science 181:175-177
46. Blakemore SJ, Rees G, Frith CD (1998) How do we predict the consequences of our actions? A functional imaging study. Neuropsychologia 36:521-529
47. O'Brien L (2003) On knowing one's own actions. In: Roessler J, Elian N (eds) Agency and self-awareness: issues in philosophy and psychology. Oxford University Press, Oxford, pp 358-382
48. Dokic J (2003) The sense of ownership: an analogy between sensation and action. In: Roessler J, Elian N (eds) Agency and self-awareness: issues in philosophy and psychology. Oxford University Press, Oxford, pp 321-344
49. Gallagher S (2008) Direct perception in the intersubjective context. Conscious Cogn 17:535-543
50. Gopnik A, Meltzoff AN (1997) Words, thoughts, and theories. MIT Press, Cambridge, MA
51. Scholl BJ, Tremoulet PD (2000) Perceptual causality and animacy. Trends Cogn Sci 4:299-309
52. Rizzolatti G, Fadiga L, GalleseV, Fogassi L (1996) Premotor cortex and the recognition of motor actions. Cognitive Brain Res 3:131-141

Affordances and the Sense of Joint Agency

J. Dokic

2.1 Introduction

All of us are aware when we are doing something. We have a sense of our own agency. We can also be aware that another agent is doing something. Thus, we have a sense of the other's own agency. The relationship between these two types of awareness of action is the subject of intense debates in the philosophy of mind and in cognitive science. Some authors argue that our awareness that we are doing something ourselves is in fact complex. It involves the *sense of agency*; we are aware of a bodily *action* in contrast to a mere passive movement. For instance, I am aware that I am slapping my hand on the desk, and not merely that my hand is slapping the desk (perhaps as the result of a reflex, or an external manipulation). But it also involves a different sense, of being the author of the action; we are aware that we *ourselves* are doing something, in contrast to another agent.

The conceptual dissociation between the sense of agency and the sense of being the author of the action raises the possibility that the sense of agency is roughly of the same kind when we are doing something and when another agent is doing something. I am aware of my action in pretty much the same way as I am aware of another agent's action. In both cases I perceive an action as such, as opposed to its being a mere passive movement. Whether or not I self-ascribe the perceived action depends on the operations of a different mechanism of self-identification [1, 2].

In this chapter, I argue for the more limited claim that the sense of agency is at least partly grounded on the perception of various kinds of affordances, both in our own case and in the case of other agents. Thus, there is a common perceptual component in our awareness that we are doing something and that another agent is doing

J. Dokic (✉)
École des Hautes Études en Sciences Sociales & Institut Jean-Nicod, Paris, France

something. Moreover, I try to show that this common component is crucial to understanding our sense of *joint* agency, when we cooperate with other agents in order to achieve a shared goal.

The claim that the sense of agency is broadly perceptual suggests that it does not involve the exercise of theoretical concepts of intentional action, as embedded in a theory of mind. If true, this suggestion can be extended to the sense of joint agency. Joint action is often pictured as a quite sophisticated achievement, which requires reflecting on the other participants' intentions. Here, in contrast, I argue for a non-mentalistic analysis of at least some genuine forms of joint action. The chapter is structured as follows. In Par. 2.2, the concept of social perception and its relationship to theory of mind or, more generally, mind-reading abilities is introduced. In Par. 2.3, the concept of affordances is defined without commitments to more radical Gibsonian claims. In Par. 2.4, a distinction is drawn between instrumental affordances (the fact that something can be done) and deontic affordances (the fact that something should be done). Paragraph 2.5 is an investigation into the neural bases of instrumental affordances, taking up the suggestion that they involve canonical neurons in the pre-motor cortex. Paragraph 2.6 distinguishes between egocentric perception of affordances (when I perceive that I can do something) and allocentric perception of affordances (when I perceive that another agent can do something). In Par. 2.7, I suggest that the perception of affordances can also be dependent on the perception of another agent's action. The perception of action can reveal affordances that would be hard or even impossible to perceive otherwise. Paragraph 2.8 introduces the concept of interpersonal affordances, which will be crucial to the proposed analysis of joint action. Finally, in Par. 2.9, I compare two models of joint action. The first model links the ability to engage in joint action to mind-reading, whereas the second model is grounded in the idea that joint action involves the participants' non-mentalizing perceptions of various kinds of personal and interpersonal affordances.

2.2
Social Perception and Mind-reading

One of the major tasks of a cognitively oriented approach to the mind is to explain our amazing skills for social interaction and understanding. A more specific question is to what extent these skills rely on a mind-reading ability, conceived as the (perhaps uniquely human) ability to understand, explain, and predict the behavior of others in terms of their mental states, including beliefs, desires, emotions, and intentions. The ability of mind-reading was initially modeled as involving the possession of a *theory of mind*. As David Premack and Guy Woodruff argued in their groundbreaking work on mind-reading [3]:

> A system of inferences of this kind may properly be regarded as a theory because such [mental] states are not directly observable, and the system can be used to make predictions about the behavior of others. (p. 515)

More recently, the mind-reading ability has been conceived more broadly as involving both *theory* and *mental simulation* [4]. Moreover, some psychologists have suggested that it is realized in an innate cognitive module encapsulated from the subject's consciously accessible knowledge [5, 6] (see Chapter 1 of [7] for a detailed review). But the main idea has been that most, if not all, mental states cannot be directly observed; rather, they must be *inferred* from the direct observation of something else, such as behavior. In this view, the human mind involves a *mind-reading system* that can take as inputs perceptual (first-order) representations and yield as outputs other (including second-order) representations (i.e., representations about other representations) that can be fed into the *practical reasoning system*, which can then produce relevant behavior, including linguistic utterances. This simple model can be called "the serial view" of the relationship between perception and mind-reading (Fig. 2.1).

For instance, I perceive my friend's avoidance behavior towards an approaching person. Reading his mind, I come to the conclusion that he does not want to meet that person (activation of the mind-reading system). I am curious to know why, and I ask him (activation of the practical reasoning system).

An objection to the serial view is that it neglects the fact that we have "social senses," to use Bernard Conein's highly appropriate phrase [8]. In other words, we can apparently *perceive* (rather than infer from anything else) social events and states of affairs, or at least we can take advantage of perceptual cues that are socially significant. As many authors have pointed out, the human mind involves a *social perception system* that directly responds to sensory stimulations and produces as outputs representations that already have a social significance:

Social perception refers to initial stages in the processing of information that culminates in the accurate analysis of the dispositions and intentions of other individuals. ([9], p. 267)
What we call "social perception" is the part of the human visual system specialized in the processing of cues of social actions and social intentions. ([10], p. 236)

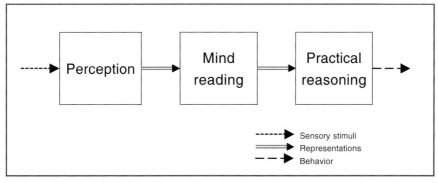

Fig. 2.1 The serial view

Thus, the concept of social perception refers to relatively low-level neural mechanisms, many of them located in the superior temporal sulcus [11, 12], that are capable of detecting socially relevant cues, such as directions of gaze, gestures, facial expressions, and speech. Pierre Jacob and Marc Jeannerod have a more restrictive concept, according to which only cues of actions and intentions directed toward conspecifics (intentions to affect the other's behavior) are included in the domain of social perception [10]. They call such actions and intentions "social," in contrast to actions and intentions directed toward inanimate objects. For our purposes in this chapter, we stick to the broader concept of social perception, according to which perception is social as soon as it is about cues of *another subject*'s actions and intentions.

The question now arises how we should modify the serial view in order to take the social perception system into account. Obviously, the outputs of the latter can serve as filtered inputs to the mind-reading system in a way that reduces the need for the construction of complex representations of socially relevant cues out of piecemeal perceptual information. Due to the computational work already done at the level of the social perception system, these cues can pop out as *sui generis* perceptual *Gestalten*. So, the simplest way of modifying the serial view remains as serial as ever (Fig. 2.2).

One might still object to the modified serial view because it does not deal with many ordinary cases of social interaction. Sometimes, if not often, the social perception system has a causal as well as a rational impact on behavior that does not seem to hinge on the mind-reading system. From a phenomenological point of view at least, we can react to the other's behavior without bringing to bear theoretical concepts of mental states, such as intentions. As Gallagher put it [13]:

> Phenomenology tells us that our primary and usual way of being in the world is pragmatic interaction (characterized by action, involvement, and interaction based on environmental and contextual factors), rather than mentalistic or conceptual contemplation (characterized as explanation or prediction based on mental contents). (p. 212)

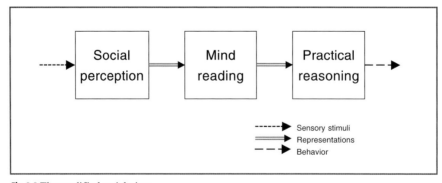

Fig. 2.2 The modified serial view

If Gallagher is right, there must be two routes from the social perception system to behavior, one of which goes through the mind-reading system but the other bypasses it. Moreover (although Gallagher may disagree on this further count), the non-mentalistic route from the social perception system to the mind-reading system can be a *representational* route, meaning that the social perception system produces as outputs conceptual (or proto-conceptual) representations that can activate the practical reasoning system, rather than having a direct impact on behavior. The upshot of this is that the subject can behave in an intelligent and flexible way on the basis of her perception of socially relevant cues. Let us call this "the parallel view" of the relationship between social perception and mind-reading (Fig. 2.3).

One might wonder how the outputs of the social perception system can be about socially relevant cues without involving theoretical concepts of mental states. But the idea of a social perception system does not entail that one perceives the relevant cues *as* social cues, in a way that would indeed mobilize concepts whose mastery depends on having a mind-reading ability. Consider, for instance, the case of biological motion. It is very plausible that, thanks to dedicated neural mechanisms, we see biological motion quite differently from any other kind of physical motion [10, 14]. Now biological motion is of course an important cue of the agent's intentions. It does not follow that we perceive biological motion *as* providing cues of intentions. In general, the concepts that figure in the output representations of the social perception system are concepts of behavioral invariants and regularities that need not be embedded in a theory of mind, but can be used in further (strictly first-order) reasoning eventually leading to appropriate behavior.

Admittedly, the parallel view is entirely schematic as it currently stands. To begin with, a detailed story should be told about the conditions under which the mind-reading system is activated. One might claim, following Gallagher [13], that this system is involved only in special cases of social observation, as opposed to more frequent cases of social interaction. There is social observation without interaction when one

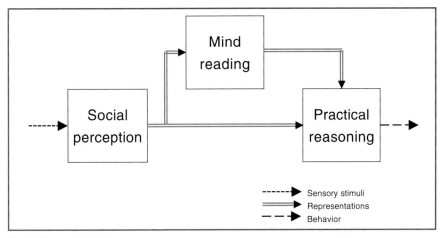

Fig. 2.3 The parallel view

observes from the outside, in a detached manner, another person doing something, or several persons interacting with each other but not with the observer. Alternatively, one might argue that the mind-reading system always attends behavior and, at least sometimes, *re-describes* the transactions between the social perception system and behavior in mentalistic terms (in line with Daniel Povinelli's reinterpretation hypothesis; see [15]). In this picture, the perception of socially relevant cues can be re-described using the conceptual resources of the mind-reading system, for instance, as cues of intentions and other mental states.

My aim here is to show that basic forms of joint action can rely on the social perception system to a considerable extent, without bringing to bear mind-reading abilities. Joint action is often deemed to involve a higher-level awareness of the other's intentions, but if I am right, this is not necessarily the case. My claim is that joint action is grounded in the perception of various kinds of affordances, including social or at least interpersonal affordances. However, before we come to the mechanisms underlying joint action, it is necessary to introduce the general concept of affordances.

2.3
The Concept of Affordances

The concept of affordances was introduced by the ecological psychologist James J. Gibson, who proposed the following definitions:

> What is meant by an affordance? A definition is in order, especially since the word is not to be found in any dictionary. Subject to revision, I suggest that the affordance of anything is a specific combination of the properties of its substance and its surfaces taken with reference to an animal. The reference may be to an animal in general as distinguished from a plant or to a particular species of animal as distinguished from other species. ([16], p. 67)
> The affordances of the environment are what it offers the animal, what it provides or furnishes, either for good or ill. ([17], p. 127)

There is little doubt that there *are* affordances in the world. From an ontological point of view, affordances can be seen as complex physical relations involving an object or a set of objects in the world and the animal's cognitive set or, more precisely, whatever physically realizes the animal's various abilities and skills (including neural mechanisms). Affordances involve various dispositional properties of the object or set of objects, namely, dispositions to elicit various types of actions or reactions on the animal's part. For instance, if the door is wide enough, it will afford the action of moving the fridge through it. If a chair has the right height, it will afford the action of sitting on it, and so on and so forth.

Affordances can be relative to a species. For instance, water counts as a standing surface for the dragonfly but not for the cat. Moreover, within the same species or even in the same individual, affordances can still be relative to the situation. As an example, a spoon can be used by a prisoner to eat his soup but also (as he suddenly realizes) to dig a tunnel in the ground that (if he is patient enough) will allow him to escape his cell.

If there are affordances, the next question is whether and how an animal can access them. Gibson's controversial claim is that affordances can be directly perceived, or at least known, without inference. For instance, one can *perceive* the spoon as affording specific types of spoon-related actions. So, perceiving affordances yields information about one's possibilities of action and reaction in one's present situation. The perception of affordances can be hard-wired or acquired through conditioning or other forms of learning.

As José Bermúdez usefully points out: "to say that affordances are directly perceived is precisely to say that instrumental relations can feature in the content of perception" ([18], p. 118). Instrumental relations relate a bodily means M (a type of bodily movement) to a goal G (a possible future state of affairs). More specifically, there is an instrumental relation when actualization of the type of bodily movement M results in G's being the case. For instance, turning the doorknob and pushing the door would lead to a specific result, namely, that the door is open. I will use the notation [M → G] to refer to instrumental relations in this sense. Thus, the claim that we can perceive affordances is the claim that the content of perception can have the form [M → G]. For instance, I perceive the doorknob and the door to which it is attached as affording a complex action, that of opening the door.

Based on this interpretation, the perception of an affordance is similar to the perception of a counterfactual state of affairs. I can perceive the affordance of the doorknob and the door even though I am not actually doing anything. What I perceive is that *if* I turned the doorknob and pushed the door in the appropriate way, the door would be open. Not all perceptions of counterfactual states of affairs are perceptions of instrumental relations. For instance, I can perceive that a particular vase is fragile, which means that *if* the vase fell to the ground, it would break. In this case, my perception is about a counterfactual state of affairs that does not involve any instrumental relation. Of course, I might *also* perceive the vase as affording a type of action, i.e., as being such that, if I pushed the vase over the edge of the table, it would fall to the ground and break. Instrumental relations are counterfactual states of affairs that are specifically about bodily means to physical goals.

Like other forms of perception, the perception of affordances can be illusory in various respects. I can have the visual experience of the doorknob and the door as affording the action of opening the door, while the door is actually locked. In such a case, my experience would not be fully veridical. It is not the case that if I turned the doorknob and pushed the door, the door would be open. On the contrary, it would stay closed, and I would need to actualize another means (probably involving the right key) to the intended goal.

The perception of affordances warrants instrumental beliefs that play a special functional role in our cognitive economy. These beliefs combine with appropriate motivational states, such as desires, to lead into action. If I perceive an affordance of the form [M → G], and if I am independently motivated to reach goal G, then I will be at least inclined to actualize the type of bodily movement M. It is very important to note that none of this involves any form of reflection on my practical reasoning. The perception of affordances is about bodily means to physical goals, but it does not require the activation of my mind-reading system.

The thesis that we can perceive affordances should be dissociated from two independent, more radical claims also endorsed by Gibson and some of his present-day followers. One of these claims is that *all* that we ever perceive are affordances. This is not an obvious implication of our previous considerations. Even if we can perceive affordances, we might also be able to perceive things that have nothing to do with our local opportunities for action. As John Campbell pointed out [19], we sometimes perceive objects, such as stars in the sky, without having the least idea of what they can be used for.

The other radical claim associated with the Gibsonian tradition is that the perception of affordances is *direct*, in the sense that we do not perceive an affordance *by* perceiving anything else (in contrast to a case in which we hear a car by hearing the sound it makes). This is controversial. For instance, Campbell argued that even if we may not be able to perceive affordances concerning other species, for instance, concrete interstices that afford nesting for pigeons, we can somehow perceive the reasons why a physical structure supports a specific affordance. As Campbell put it, "we see the ground of the affordance" ([19], p. 143). In general, it might be argued that the perception of an affordance is always indirect, in the sense that we perceive opportunities for various types of action *by* perceiving categorical properties, such as shape, size, and texture, of a particular object, surface, or structure.

In what follows, I use the concept of affordances independently of the more radical claims that we perceive nothing but affordances and that we perceive them directly. The general idea that we can perceive affordances, or at least know about them from sensory experience without inference, even though our perception of affordances is typically if not always based on the perception of something other than affordances, is all I need for present theoretical purposes.

2.4
Instrumental *vs* Deontic Affordances

An assumption of the foregoing definition of affordances is that they are motivationally "cold," in the sense that their perception need not be accompanied by any strong inclination to realize them. Perceiving an affordance of the form [M → G] can ground a belief of the form "I *can* do G by M-ing." The perceiver can thereby know, on the basis of observation, that *if* she were independently motivated to do G, she would be inclined to do G by M-ing (for instance, to drink by manipulating the glass in the appropriate way). We perceive many instrumental affordances around us, but fortunately we are seldom if ever inclined to act according to all of them.

Now perhaps there is another kind of affordance whose perception is motivationally "warm," in the sense that it necessarily involves some inclination to realize them.

Let us say that these affordances are *deontic* rather than instrumental. Like instrumental affordances, their perception can ground a belief of the form "I can do G by M-ing," but unlike instrumental affordances, they can also ground a belief of the form "I *should* do G by M-ing," which, we may suppose, involves the motivation to actualize the bodily movement M in order to reach the goal G.

Perhaps deontic affordances are perceived in emergency situations, especially when the latter have a strong moral relevance. For instance, I see from the shore that a person is drowning and I immediately jump into the water to try and save her. What I perceive is not merely an instrumental affordance. Of course, I know by observation that I can save the person by actualizing a series of appropriate movements, but I also realize that I *should* save the person and, if I am not morally insane, I immediately act accordingly.

There are two possible views about the relationship between instrumental and deontic affordances. According to one view, the perception of deontic affordances is just the perception of instrumental affordances accompanied by independent motivational states; for instance, a desire to save the person. In the second view, the perception of instrumental affordances is the perception of deontic affordances that have been somehow *inhibited*. One syndrome that could be relevant to assess these views is *utilization behavior*, observed in patients with a bilateral focal frontal lesion [20, 21]. Here is how Tony Marcel described this syndrome [21]:

> If there is some object that can be used or manipulated within the patient's vision and within reach, the patient will use it to perform actions appropriate to the object, though they have been asked not to do so. [The patients] cannot stop themselves performing actions with the irrelevant object. (p. 77)

For instance, the patient sees a pair of glasses lying on the table, and cannot stop putting them on him, even if he is already wearing another pair. Marcel observed that in utilization behavior the abnormal actions are "environment-driven." In our terminology, they are driven by the perception of deontic affordances that the patients cannot inhibit, perhaps because of their frontal lesions. One may speculate that such patients lack the executive resources to transform their "warm" perceptions of deontic affordances into "cold" perceptions of merely instrumental affordances. If this is right, then the perception of deontic affordances is prior to the perception of instrumental affordances in the order of explanation, in the sense that the latter should be considered as a suppressed form of the former.

2.5
Canonical Neurons as Reflecting Instrumental Affordances

In a series of single-neuron recording experiments on macaque monkeys, Giacomo Rizzolatti and his colleagues in Parma investigated the functional properties of neurons in area F5, the rostralmost sector of the ventral premotor cortex that controls hand and mouth movements [22]. A fundamental functional property of area F5 is that most of its neurons do not discharge in association with elementary movements but are active during purposeful object-oriented actions, such as grasping, tearing, holding, or manipulating objects. Although the majority of neurons in F5 are purely motor neurons, area F5 also contains two classes of visuomotor neurons: *canonical neurons* and *mirror neurons*. The latter class of neurons will be discussed below; let us now focus on the former class.

Canonical neurons are activated during the execution of goal-related movements and also discharge during object observation, typically showing congruence between the type of grip they motorically code and the size/shape of the object that visually drives them. Since they associate a motor program with a perceived object, it is reasonable to think of them as reflecting affordances [23, 24]. Here is Susan Hurley's speculation about the origin of canonical neurons' functional properties [24]:

> It could be predicted that cells that mediate the association between copies of motor signals and actual input signals might come to have both motor and sensory fields. Suppose an animal typically acts in a certain way on the perceived affordances of a certain kind of object: eating a certain kind of food in a certain way, for example. There will be associations between copies of the motor signals for the eating movements and a multimodal class of inputs associated with such objects and the eating of them. Any cells that mediate this association might thus have both sensory and motor fields that between them capture information about the affordances of the objects in question. Canonical neurons are candidates for such predicted sensorimotor affordance neurons. (p. 235)

Note that Hurley's description depends on the assumption that the animal has often realized in the past the affordances it has perceived–otherwise the sensorimotor associations she is talking about would not have arisen. This assumption is consonant with the conclusion reached in the previous paragraph, about the explanatory priority of the concept of deontic affordances over that of instrumental affordances. More precisely, we can assume that the perception of instrumental affordances always has some deontic and thus motivational component, which can be more or less repressed depending on what is practically relevant in the situation. To sum up, canonical neurons can be seen as at least contributing to the neural basis of our ability to perceive basic instrumental affordances, namely, those that concern hand and mouth transitive actions.

2.6
Egocentric *vs* Allocentric Perception of Affordances

Personal affordances are relative to a particular agent's dimensions, abilities, and skills. Their perception can be called *egocentric* when the agent is the perceiver herself. When I have an egocentric perception of an affordance, I have reason to believe that *I* can do something. In contrast, perception of personal affordances can be called *allocentric* when it concerns *another* agent's dimensions, abilities, and skills. When I have an allocentric perception of an affordance, I have a reason to believe that *someone else* can do something.

Do we actually have the ability to perceive the world as affording someone else's actions? An affirmative answer is suggested by experiments conducted by Thomas Stoffregen [25]. Subjects were asked to estimate both the *maximal* and the *preferred* heights of an adjustable seat relative to actors of different sizes. In one condition, the subjects' estimations were based on the perception of the actor standing still next to the seat. In another condition, they were based on the perception of cinematic information in the absence of the actor (but in the presence of the seat). The authors found that the subjects' estimations were by and large correct. These results suggest that the subjects perceive personal affordances that can be specified by statements such as "This chair is almost too high for this person to sit on" (in the maximal height estimation condition) and "This chair is high enough for this person to sit on comfortably" (in the preferred height estimation condition).

There are two important differences between egocentric and allocentric perceptions of affordances. First, even though my egocentric perception of an affordance of the form [M → G] warrants the belief that I can do G by M-ing, the self need not be explicitly represented *in perception*. I may perceive the reachability (by me) of the apple, even though I am not a component of the visual field, unlike the apple (see [26]). In contrast, allocentric perception of affordances *requires* explicit representation of the relevant agent, and thus involves slightly more complex representational resources. Second, as we have seen above, my egocentric perception of an affordance of the form [M → G] will normally lead to action in concert with the desire to do G. In contrast, my allocentric perception of an affordance warrants the belief that the other can do G by M-ing, but this won't necessarily lead to action in concert with the desire to do G. Think of a case in which I am unable to actualize the type of bodily movement M.

It follows that our ability to perceive affordances allocentrically cannot be explained simply in terms of congruence with our own motor programs. Of course, the allocentric perception of something that another person can do is sometimes accompanied by the egocentric perception of something that I can do too. In such a case, I perceive the world as affording an action that either I or the other person can do. As Stoffregen's experiments showed, though, this cannot be the general case. Even if we assume that the allocentric perception of an affordance involves the covert simulation of an action of the same type as the action that the other can do in her situation (namely, activation of a broadly congruent motor program), the subjects' estimations in these experiments are clearly too fine-grained to be deduced from the stimulation process alone.

It remains the case that allocentric perception of affordances does not require mind-reading abilities. Perceiving that another person can reach a physical goal G by actualizing a type of bodily movement M falls short of ascribing mental states, such as intentions, to that person. All that is needed is the first-order ability to represent other people and their bodily movements (as well as counterfactual relations).

2.7
Mirror Neurons and Action-dependent Affordances

What I want to suggest now is that the perception of another subject's individual *action* also reveals new personal affordances, which otherwise would be unperceived or at least harder to experience.

As we have seen, area F5 of the ventral premotor cortex contains another class of visuomotor neurons, "mirror neurons." These fire both when the subject makes a goal-directed action, such as grasping a nearby object, and when the subject observes a similar action performed by another agent. In contrast to canonical neurons, mirror neurons do not fire in the mere perceptual presence of the object [27, 28].

In a famous, mentalistic interpretation ([29, 30], mirror neurons constitute the neural basis of our ability to *understand* the other's intentions, such as the fact that the other has a certain goal G that he expects to attain by actualizing the bodily movement M. Mirror neurons, and more generally the mirror systems to which they belong, serve a *retrodictive* function; their role is to reconstruct the agent's intention from an observation of her bodily movements. The leading idea is that motor resonance allows the observer to know what the other intends to do because "he knows[s] its outcomes when he does it" ([31], p. 396). If this is the case, mirror neurons are central components of the mind-reading system, at least as far as intentions are concerned.

An objection to the mentalistic interpretation of mirror neurons is that the latter's function is not to produce a representation of the goal from the observation of bodily movement but rather to predict or anticipate the movement from a *prior*, independently constructed representation of the goal outside the motor system [32]. At least two empirical considerations seem to support this objection. First, although mirror neurons do not discharge when the subject observes a *pretend* transitive action (for instance, the experimenter acts as if he is grasping an object where there is none), they may discharge when a monkey watches the experimenter *about* to grasp an unseen object (hidden behind a screen) that the monkey independently knows to be there and edible. In such a case, it seems clear that the action's goal is not perceptually given.

Another consideration is that mirror neurons can depend on a more general "action plan" [33]. The same action (for instance, grasping an object) can activate different mirror neurons depending on the action to be made at the next step (eating the object or placing it in a container). According to Gergely Csibra, "[t]his is a clear demonstration the [mirror neurons] take into account the further goal, and not just the perceived action, when responding to observed actions" ([32], p. 445).

Independently of the foregoing objection to the mentalistic interpretation of mirror neurons, one may also argue that mirror neurons and systems embody information about the observer's *own* action opportunities, rather than generate (by themselves) a representation of the other agent's intention. In our terminology, mirror neurons underlie *egocentric* perceptual representations of personal affordances, i.e. of what the observer herself can do in a given situation. Günther Knoblich and Scott Jordan formulated this argument [34]:

The kind of action understanding the mirror system provides is ego-centered and does not necessarily include an explicit representation of another agent. As a consequence, organisms endowed with a mirror system may have the ability to understand that objects are affected in a way that in which they could also affect them, but they may not understand that the peer who is producing the action is an agent like themselves. (p. 116)

For instance, let us assume that a monkey observes a conspecific picking up a red berry in a bush. What the monkey perceives is not the other's intention to pick up a red berry, but rather a specific instrumental relation: it sees that the actualization of a complex bodily movement of a given type (stretching the arm in a given direction, grasping the berry, and so on) would lead to an interesting outcome (picking up more berries, assuming that there are some left to eat in the bush). In this interpretation, mirror neurons are like canonical neurons in that they can be part of a perceptual system capable of revealing personal affordances *egocentrically*. They are not redundant since they can reveal affordances that would be more difficult to perceive by way of canonical neurons in the absence of the observed action. Once again, the observation of the other monkey picking up a red berry in a bush might make the observer realize that there are edible berries in the bush to be picked up in certain way, where these berries would otherwise be barely visible.

Rizzolatti *et al.* [28] anticipated such an interpretation in the following passage:

When the monkey observes another monkey grasping a piece of food, the obvious action to take would be, for example, to approach the other monkey, but certainly not to repeat the observed action. (p. 667)

However, it is not clear that there is a real contrast here. When there are enough berries in the bush, it might be better for the monkey to wait until the other monkey has left and directly approach the bush, rather than taking the risk of stealing the berry already in the hands of its conspecific.

Knoblich and Jordan's interpretation of the function of mirror neurons is plausible (see [35]), although two caveats are perhaps in order. First, the crucial issue is whether the activation of mirror neurons should be conceived as (part of) an exercise of mind-reading, not whether it involves the perception of an agent as such. According to a non-mentalistic interpretation of the function of mirror neurons, my perception of someone else involved in an action reveals a *new* personal affordance, namely, that a certain goal can be achieved by way of a certain type of bodily movement (whose occurrence I am currently seeing). This is compatible with the fact that I explicitly represent the other as an agent, at least in the minimal sense according to which an agent is the locus of biological motion quite unlike mere physical movement. In this sense, I can explicitly represent an agent without representing her as having mental states such as intentions, i.e. without mobilizing my mind-reading system.

The second caveat is that the perception of someone else's action can also reveal personal affordances *allocentrically*, that is, as affordances relative to the observed agent herself. When the relevant bodily movement can be actualized either by the observer or the observed agent, the difference between egocentric and allocentric perception of the affordance is not obvious. (Of course, if G is the goal of picking up and eating a *particular* berry, then the more the monkey's action unfolds in time, the less I can perceive an affordance of the form [M → G] *egocentrically*, since as soon as the monkey reaches G, that goal is no longer available to me.) What is important is that in neither case does the observer have to mobilize theoretical concepts of mental states. Whether the perception of the affordance is egocentric or allocentric, it falls short of *ascribing* the goal to anyone, in the form of an *intention*. It merely contributes to revealing to the observer a goal that can be reached in a certain way.

In a nutshell, the suggestion is that there are two kinds of observable personal affordances. Some of them are *action-independent*, in the sense that they can be perceived without perceiving any action, whereas others are *action-dependent*, in the sense that they can be perceived only by perceiving an action. Just like action-independent affordances, action-dependent affordances can be perceived either egocentrically (relative to oneself) or allocentrically (relative to the observed agent). In neither case need the action be mentalistically conceived by the observer.

What remains controversial is the role of mirror neurons in underlying perceptions of action-dependent affordances. If Csibra's interpretation is on the right track, mirror neurons cannot generate by themselves representations of goals, which must be represented outside the motor system. Now, given a sufficiently liberal conception of perception, one might allow for the perception of an action-dependent affordance of the form [M → G] in a situation in which the only observed action involves the subgoal G' rather than G itself. For instance, by observing the action of grasping an object, one might perceive that there is a bodily means to the goal of it being eaten, or alternatively of it being placed in a container. In other words, the fact that no action involving the goal G is perceived does not entail that G is not perceptually represented (as a possible state of affairs). Features of the situation and the observer's own perceptual history and expectations might be rich enough to enable the perception that there is a bodily means [M → G]. This perception, assuming that it is available to us, cannot be based on motor mirroring only, as the goal G must be represented independently, outside the motor system (for a more general criticism of the motor theory of social cognition, see [36]).

2.8
Interpersonal Affordances

What we have studied so far are *personal* affordances, either relative to oneself or to someone else. Personal affordances can be contrasted with *interpersonal* affordances, which are relative to at least two subjects. Interpersonal affordances are opportunities for *joint* action, and their perception has both an egocentric and an allocentric aspect.

The questions arise of when the world affords joint action of a given type, and when such an affordance can be perceived by the participants. Michael Richardson and colleagues [37] investigated cases in which basic joint actions occur spontaneously, i.e., without prior planning. They hypothesized that the transition points between solo and duo actions are determined by the subjects' perceptions of relevant interpersonal affordances, determined by complex relations between the constraints provided by the current task, the environment, and their own dimensions and abilities. Thus, they compared the perception and actualization of intrapersonal, interpersonal, and tool-based affordances. In one series of experiments, subjects were asked to *judge* whether they would grasp wooden planks using one hand, two hands, with a special tool that extended their reach, or with the help of another person. In another series of experiments, they were asked to actually *grasp* the planks using one of these methods. The authors found that the participants either judged that they would switch, or actually switched, between the different modes of grasping, at transition points occurring at similar "action-scaled ratios," conceived as complex relations between the subjects and their environment:

> Cooperating individuals come together to actualize interpersonal affordances in much the same way as two limbs come together to actualize intrapersonal affordances. The similitude between affordances at multiple levels of the animal–environment system–the body, the body-tool, and the body-body–is being suggested here, where the emergence of cooperation and coordination at each level (both intrapersonal and interpersonal) is a result of the same intrinsic informational constraints. That is, despite the intuition that cooperative action is substantially different from solo action, an understanding of cooperative activity in terms of affordances suggests that there is a similarity in how joint and solo activity is constrained and organized. ([37], p. 847)

When an agent uses a familiar tool in order to realize an affordance, the tool becomes a functional part of the agent's action system, just as biological bodily parts. The agent's action system has been *extended* with the use of the tool. In a similar fashion, the agent can use another's action capabilities to extend her action system. As Richardson et al. ([37] p. 856) put it, "body-tool and social action systems can be understood and studied as a single synergy or effectivity" (see also [38]).

These findings are relevant to the issue of whether mind-reading is involved in joint cooperative action, and, if so, to what extent. One can argue that just as mind-reading is not essentially involved in tool-based activity, it need not be involved in joint cooperative action either. It certainly does not follow that when I engage in a joint action, I perceive the other as a mere tool (violating Kant's maxim that the other should not be treated as a means to another end). On the contrary, I directly perceive the other as a person, or at least as an autonomous biological agent. But, as we have seen above, this perception is independent of mind-reading, conceived as the sophisticated ability to ascribe propositional mental states like beliefs, desires, and intentions.

In the simple case involving two agents, an interpersonal affordance has the form $[(M1 + M2) \rightarrow G]$, which means that the conjunction of a bodily movement of type M1 and a bodily movement of type M2 would lead to goal G. When I perceive such an affordance, I have a reason to believe that *we* can do G, provided that I actualize a bodily movement of type M1 and you actualize a bodily movement of type M2.

These bodily movements are sometimes reversible, but need not be. Perhaps only you can make a bodily movement of type M2. Analogously, when you perceive this affordance, you have a reason to believe that *we* can do G, provided that you actualize a bodily movement of type M2 and I actualize a bodily movement of type M1. For instance, we both perceive that we can lift this (visually presented) wooden plank if one of us grasps the plank by one end while the other grasps it by the other end.

Now if I perceive that you are *about* to actualize a bodily movement of type M2, for instance, that you are ready to grasp the wooden plank by one of its ends, I can perceive a new *personal* affordance, that a bodily movement of type M1 would be enough to lead to our joint goal G. Analogously, if you perceive that I am *about* to actualize a bodily movement of type M1, you perceive a new affordance of the form [M2 → G]. Of course, in many cases, my bodily movement actually depends on the development of yours and vice versa (think of a complex dancing situation), which means that our perceptions of the relevant personal affordances have to occur at about the same time.

2.9
Two Models of Joint Action

Some philosophers, among them Michael Bratman [39] figures prominently, have claimed that joint action depends on the participants having second-order intentions, i.e., intentions about the other participants' intentions. In this view, the ability to engage in joint action requires a quite sophisticated mind-reading system. In a similar vein, Michael Tomasello and his collaborators wrote about joint cooperative activities that "the goals and intentions of each interactant must include as content something of the goals and intentions of the other" ([40], p. 680). In this paragraph, I explore further the alternative view that the ability to engage in joint action can be explained, at least to some extent, without reference to the mind-reading system, in terms of the perception of personal and interpersonal affordances. More precisely, if joint action involves the manipulation of perceptual and non-perceptual representations of various kinds of affordances, these representations remain first-order, i.e., they do not explicitly represent mental representations (like intentions) as such.

According to Bratman's influential model [39], joint action involves a collective intention or "we-intention," which can be analyzed as follows (once again sticking to the case of two agents):

Bratman's Model
We intend to make it the case that G if and only if:
B1. I intend to make it the case that G.
B2. You intend to make it the case that G.
B3. Each participant intends to make it the case that G in accordance with and because of B1 and B2, and meshing sub-plans of B1 and B2.

Clauses B1 and B2 ensure that both participants have the same goal (for instance, moving a wooden plank from one place to another). B3 attributes to each participant a reflexive intention, which is explicitly about or represents the other participant's intention. Bratman postulated reflexive, interlocking intentions, which require mind-reading abilities, mainly in order to rule out what he calls the "Mafia sense" of "we're doing G together":

> You and I intend that we go to New York together; and this is common knowledge. However, I intend that we go together as a result of my kidnapping you, throwing you in my car, and thereby forcing you to join me. The expression of my intention, as we might say, is the Mafia sense of "we're going to New York together." ([39], p. 118)

Clause B3 is supposed to guarantee that we are going to New York together in a way in which your intention plays a causally efficacious role in our joint action, i.e., to ensure that we have "mutually noncoerced intentions in favor of the joint activity" ([39], p. 108).

That the participants' intentions should be noncoerced in this sense is of course a central requirement for there being a genuine joint action. The question is whether the attribution of reflexive intentions is the only way to meet this requirement. Here I would like to suggest that we can rule out the Mafia sense of "We're doing G together" without ascending to a higher-order level of intentions, by bringing into the picture the participants' perceptions of personal and interpersonal affordances. A tentative model of joint action that parallels Bratman's main clauses without postulating interlocking reflexive intentions can be formulated:

The Affordance Model
We intend to make it the case that G if and only if:
A1. I intend to make it the case that G.
A2. You intend to make it the case that G.
A3. Each participant perceives an interpersonal affordance of the form [(M1 + M2) → G].
A4. I perceive that you are about to make bodily movement M2, and you perceive that I am about to make bodily movement M1.
A5. On the basis of A4, I perceive the personal affordance [M1 → G] and you perceive the personal affordance [M2 → G].
A6. On the basis of A1 and A5, I intend to make it the case that G via M1 and you intend to make it the case that G via M2.

The affordance model of joint action has two main advantages over Bratman's model. First, it does not ascribe any reflexive intentions to the participants of the joint action. The causal efficacy of our respective intentions is already taken care of at the level of perception. In particular, my egocentric perception of the personal affordance [M1 → G] depends on my allocentric representation of you being about to realize the personal affordance [M2 → G], and vice versa. Our manifest inclinations to actualize bodily movement of relevant types *reveal* new personal affordances to each of us. For simplicity's sake, I have ignored the fact that the action might involve sub-plans, but these can be accounted for in the spirit of the affordance model, namely, by introducing intermediary interpersonal and personal affordances, about sub-goals.

Second, several critics of Bratman's model [41-43] have independently objected to the intelligibility of ascribing to a person an intention whose content refers to another person's intention, arguing that such a content cannot be *under one's control*. In general, I cannot directly control the other's intention; otherwise it would not be her *intention*. The only intentions ascribed to the participants of the joint action according to the affordance model are, initially, general intentions to reach a certain goal (clauses A1 and A2) and, eventually, more specific intentions to reach that goal in a certain way (clause A6). In fact, the affordance model need not even ascribe to the participants of the joint action final intentions with a *shared* content, since M1 and M2 might be different types of bodily movements.

If the affordance model is on the right track, basic forms of joint action can bypass the mind-reading system, and thus are potentially available to "mind-blind" creatures. Indeed, this model might yield a good enough description of joint activity to be found in creatures lacking a theory of mind, such as young children and non-human animals (consider, for instance, group hunting in lions).

Of course, there is no question that mind-reading greatly *enhances* the ability to engage in joint action, by enabling more complex and controlled forms of cooperation. Bratman himself added to his analysis of joint action a "common knowledge" clause ensuring that the participants know that clauses B1–B3 are satisfied. Although this is controversial, it is quite possible that this additional clause, or the analogue clause that A1–A6 are satisfied in the case of the affordance model, corresponds to a level of self-awareness that requires mind-reading abilities. Indeed, human joint action often involves levels of *mutual support*, *error correction*, and *role reversal* that are not observed in non-human animals [44], although some of these features might just require more perceptual flexibility.

Two important conclusions are still in order. First, basic forms of joint action, even in humans, do not require the mind-reading system. Second, more sophisticated forms of joint action, which clearly do require the mind-reading system, might still have a perceptual basis as described by clauses A1–A6 of the affordance model. Our sense of acting together has a strong perceptual component.

2.10
Conclusions

This chapter investigated our *sense of joint agency*, conceived as the perceptual sense that we are acting together. I have argued that a central component of our awareness that we are cooperating in order to achieve a shared goal is our egocentric and allocentric perceptions of various kinds of personal and interpersonal affordances.

I have discussed several ways in which the mere perception of another person modifies our perception of what can be done, either individually or jointly (see also [45]). First, in Stoffregen's experiments [25], the perception of a standing actor is enough to make us perceive what can be done *by him*, for instance sitting comfortably on a given seat. Second, in Richardson's experiments [37], the perception of another agent makes available new affordances, concerning what can be done *by us* and, if the other is actually cooperating, what *I* can do given my now suitably extended action system. In general, the perception of another agent's action reveals new personal affordances, either relative to the other or to myself.

I have also tried to show how these various kinds of affordances can be exploited in joint action, and more generally how their perception can have a sophisticated impact on behavior, via situation- and body-reading rather than mentalizing. Thus, a generalized theory of affordances can contribute to explain the intelligibility and rationality of at least some of our social behavior without relying too much on mind-reading conceptual resources.

References

1. Dokic J (2003) The sense of ownership: an analogy between sensation and action. In: Roessler J, Eilan N (eds) Agency and self-awareness. Issues in philosophy and psychology. Clarendon Press, Oxford
2. Jeannerod M, Pacherie E (2004) Agency, simulation and self-identification. Mind and Language 19:113–146
3. Premack D, Woodruff G (1978) Does the chimpanzee have a theory of mind? Behavioral and Brain Sciences 1:515-526
4. Perner J (1996) Simulation as explicitation of predication-implicit knowledge about the mind: arguments for a simulation-theory mix. In: Carruthers C, Smith PK (eds) Theories of theories of mind. Cambridge University Press, Cambridge
5. Baron-Cohen S, Leslie A, Frith U (1985) Does the autistic child have a "theory of mind"? Cognition 21:37-46
6. Leslie A (1987) Pretence and representation: the origins of a "theory of mind." Psychol Rev 94:412-426
7. Goldman AI (2006) Simulating minds. The philosophy, psychology and neuroscience of mindreading. Oxford University Press, Oxford
8. Conein B (2006) Les sens sociaux. Trois essais de sociologie cognitive. Economica, Paris
9. Allison T, Puce A, McCarthy G (2000) Social perception from visual cues: role of the STS region. Trends Cogn Sci 1:267-278
10. Jacob P, Jeannerod M (2003) Ways of seeing. The scope and limits of visual cognition. Oxford University Press, Oxford
11. Perrett DI, Smith PAJ, Potter DD et al (1985) Visual cells in the temporal cortex sensitive to face view and gaze direction. Proc R Soc London B 223:293–317
12. Perrett DI (1999) A cellular basis for reading minds from faces and actions. In: Hauser M, Konishi M (eds) Neural mechanisms of communication. MIT Press, Cambridge, MA
13. Gallagher S (2005) How the body shapes the mind. Oxford University Press, Oxford
14. Johansson G (1973) Visual perception of biological motion and a model for its analysis. Percept Psychophys 14:201-211
15. Penn DC, Holyoak KJ, Povinelli DJ (2008) Darwin's mistake: explaining the discontinuity between human and nonhuman minds. Behav Brain Sci 31:109-178
16. Gibson JJ (1977) The theory of affordances. In: Shaw RE, Bransford J (eds) Perceiving, acting, and knowing: toward an ecological psychology. Erlbaum, Hillsdale, NJ

17. Gibson JJ (1986) The ecological approach to visual perception. Erlbaum, Hillsdale, NJ (Original work published 1979)
18. Bermúdez J (1998) The paradox of self-consciousness. MIT Press, Cambridge, MA
19. Campbell J (2002) Reference and consciousness. Oxford University Press, New York
20. Lhermitte F (1983) "Utilization behaviour" and its relation to lesions of the frontal cortex. Brain 106:237-255
21. Marcel T (2003) The sense of agency: awareness and ownership of action. In: Roessler J, Eilan N (eds) Agency and self-awareness. Issues in philosophy and psychology. Clarendon, Oxford
22. Rizzolatti G, Carmada R, Gentilucci M et al (1988) Functional organization of area 6 in the macaque monkey. II Area F5 and the control of distal movements. Exp Brain Res 71:491-507
23. Miall RC (2003) Connecting mirror neurons and forward models. NeuroReport 14:2135-2137
24. Hurley S (2006) Active perception and perceiving action: the shared circuits model. In: Gendler TS, Hawthorne J (eds) Perceptual experience. Clarendon, Oxford
25. Stoffregen TA, Gorday KM, Sheng YY et al (1999) Perceiving affordances for another person's actions. J Exp Psychol Hum Percept Perform 25:120-136
26. Perry J (1993) Thought without representation. In: The problem of the essential indexical, and other essays. Oxford University Press, Oxford
27. Rizzolatti G, Fadiga L, Matelli M et al (1996) Localization of grasp representations in humans by PET. 1. Observation versus execution. Exp Brain Res 111:246-252
28. Rizzolatti G, Fogassi L, Gallese V (2001) Neurophysiological mechanisms underlying the understanding and imitation of action. Nat Rev Neurosci 2:661-670
29. Gallese V, Goldman AI (1998) Mirror neurons and the simulation theory of mind-reading. Trends Cogn Sci 12:493-501
30. Gallese V (2003) The manifold nature of interpersonal relations: the quest for a common mechanism. Phil Trans Royal Soc London B 358:517-528
31. Gallese V, Keysers C, Rizzolatti G (2004) A unifying view of the basis of social cognition. Trends Cogn Sci 8:396–403
32. Csibra G (2009) Action mirroring and action understanding: an alternative account. In: Haggard P, Rosetti Y, Kawato M (eds) Sensorimotor foundations of higher cognition. Attention and Performance XXII. Oxford University Press, Oxford
33. Fogassi L, Ferrari PF, Gesierich B et al (2005) Parietal lobe: from action organisation to intention understanding. Science 308:662–667
34. Knoblich G, Jordan JS (2002) The mirror system and joint action. In: Stamenov MI, Gallese V (eds) Mirror neurons and the evolution of brain and language. John Benjamins, Amsterdam
35. Pacherie E, Dokic J (2006) From mirror neurons to joint actions. Cogn Sys Res 7:101–112
36. Jacob P, Jeannerod M (2005) The motor theory of social cognition: a critique. Trends Cogn Sci 9:21-25
37. Richardson MJ, Marsh KL, Baron RM (2007) Judging and actualizing intrapersonal and interpersonal affordances. J Exp Psychol Hum Percept Perform 33:845–859
38. Marsh KL, Richardson MJ, Baron RM et al (2006) Contrasting approaches to perceiving and acting with others. Ecol Psychol 18:1–37
39. Bratman ME (1999) Faces of intention. Selected essays on intention and agency. Cambridge University Press, Cambridge
40. Tomasello M, Carpenter M, Call J et al (2005) Understanding and sharing intentions: the origins of cultural cognition. Behav Brain Sci 28:675–735
41. Baier AC (1997) Doing things with others: the mental commons. In: Alanen L, Heinämaa S, Wallgren T (eds) Commonality and particularity in ethics. Open Court, London

42. Stoutland F (1997) Why are philosophers of action so anti-social? In: Alanen L, Heinämaa S, Wallgreen T (eds) Commonality and particularity in ethics. Open Court, London
43. Velleman D (1997) How to share an intention. Phil Phenom Res 62:29-50
44. Livet P (1994) La communauté virtuelle. Éditions de l'Éclat, Combas
45. Sebanz N, Bekkering H, Knoblich G (2006) Joint action: bodies and minds moving together. Trends Cogn Sci 10:70-76

Section II
Brain, Agency and Self-agency: Neuropsychological Contributions to the Development of the Sense of Agency

The Neuropsychology of Senses of Agency: Theoretical and Empirical Contributions

M. Balconi

3.1
Different Types of the Sense of Agency

This chapter considers the two different levels of agency: one comprising lower-level, pre-reflective, and sensorimotor processes (*feelings*) and the other higher-order, reflective, or belief-like processes (*judgments*). Here, different theoretical and methodological perspectives are adopted in order to represent a compound view of the sense of agency. In addition, the concepts of "minimal" and "narrative" self are analyzed, both of which contribute to the individual's identity. As suggested by recent models, short-term and long-term representations of agency are needed to explain the contribution of experiences and actions to the construction of the subjective sense of continuity along one's personal story. In this perspective, agency is represented as the *present sense of self in action*, as well as the *continuous sense of self in existence*.

The sense of self can be specified as not merely an awareness of the self with respect to actions but also an awareness of these actions as being one's own. Proprioception is an example of the first usage whereas in the second one we mean the sensations, thoughts, intentions, and phenomenal experience recognized from within and known only to me. This knowledge is neither inferential nor observational. In a third usage, the self can be seen as a channel of information that informs us about ourselves but not about the world or about others, i.e., it is a kind of knowing that is privileged (*self knowledge*), transparent, necessarily veridical, and not vulnerable to error [1].

Recent conceptual developments have distinguished between an implicit level of "the feeling of agency" from an explicit level of "the judgment of agency" [2]. As noted above, the first is characterized by lower-level, pre-reflective, sensorimotor

M. Balconi (✉)
Department of Psychology, Catholic University of Milan, Milan, Italy

processes, and the second by higher-order, reflective, or belief-like processes. Sensorimotor processes that characterize the feeling level may run outside of consciousness (but may be available to awareness). This is supported by empirical evidence in which, for example, minor violations of intended actions or action consequences (i.e., brief temporal delays in sensory feedback) do not necessarily enter awareness, while neural signatures of such violations can be observed.

Experimental operationalization of the sense of agency must consider the distinction between these different levels of agency and thus engage in systematic explorations of the multiple indicators of agency and their possible interplay. Nevertheless, empirical investigations often focus on judgments or attributions of agency involving subjective reports and thus, potentially, of errors through misidentification. By contrast, multivariate approaches that include implicit measures (kinematics, eye movements, motor potentials, brain activity, etc.) may also tap into the *feeling level* of agency, which allows for an integrated view on the sense of agency.

Besides the two above-mentioned levels of agency, a third level can be proposed. It is related to an attributive process, which can be considered as a higher conceptual level of representing the self-in-action and which also includes the individual representation of agency in a social context. The *sense of a moral responsibility* is directly related to this high-level sense of agency. In ascribing moral responsibility to the sense of agency, we must include specific criteria, such as the presence of an internal plan of action, including a representation of a specific behavior and sufficient insight into the normally possible consequences of that behavior. The agency system, carrying out the behavior is embedded in a normative system that evaluates behaviors according to a normative rule as acceptable or not acceptable. Normative expectations are internalized and individually adopted normative rules that offer a standard of evaluation of a person's actions. Since a normative rule is determined by culturally mediated social interactions, responsibility is an essential culture-dependent phenomenon [3]. The competence to decode another's action intentions is not the only necessary element for ascribing responsibility; rather, the socio-normative dimension demands an additional component: the understanding of norms and expectations and the capacity to act accordingly. Thus, several additional capacities are required, such as the representation of shared norms about what is expected, the ability to detect deviations from those norms, and to do this not only from a first-person perspective but also from a third-person perspective [3].

The applicability of this threefold categorization of the sense of agency has been explored in the clinical setting, more recently with particular focus on the moral level. Socio-normative behavior may break-down in different ways in patients with psychopathological deficits or neurological lesions. For example, whereas a patient with a specific lesion in the prefrontal cortex may be able to represent a shared norm and to detect deviations but fails to correct his behavior accordingly, another patient, with a lesion located in an another prefrontal site, may not even be able to represent the shared norm.

The meta-representational dimension of social interaction based on social standards and normative judgments has to be distinguished from the subjective cognitive dimension of feeling and judgment, in that the latter are ontogenetically and phylogenetically more basic components. From an ontogenetic point of view, feeling and judgment do not require meta-representations, which therefore obviates the need for a model. Phylogenetically, they do not require the construction of socio-cultural normative rules. Nevertheless, the individual cognitive dimension and the socio-normative dimension constantly interact and re-model each other by bottom-up and top-down processes. For example, based on our meta-representational normative judgments and the standards of our society, we may alter our judgments about our actions: if a certain action is socially highly accepted, we are more inclined to attribute agency to ourselves.

3.2
Feeling and Judgment in the Sense of Agency

The *feeling of agency* has been described as implicit, running outside awareness but available to our conscious awareness. In this first level of agency, visual-motor incongruence is sometimes registered at the neural level (extrastriate and posterior parietal cortices) but does not necessarily enter awareness, leading to a correct judgment of the feedback as incongruent. A neural response towards sensorimotor incongruities of which subjects are not explicitly aware has been reported [4]. By contrast, in the second-level of agency, judgment has been described as being of a reflective and attributive nature, informed by conceptual thoughts. There is evidence that the prefrontal cortex is required at the level of conscious monitoring but not at the level of sensorimotor integration [5] (Fig. 3.1).

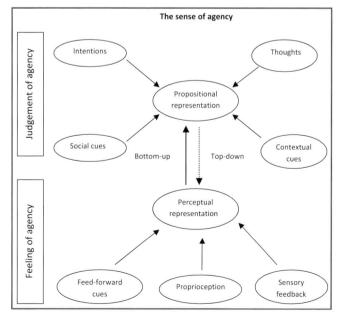

Fig. 3.1 The two-step account of agency

Another interesting bipartition with respect to agency is between *detached* and *immersed awareness*, a distinction that may be understood as being between "me" and "I." Immersed awareness is the kind of non-reflective experience one has when fully engaged in an activity, while detached awareness requires a form of reflective consciousness, in which the agent mentally observes himself acting. Generally, detached experience includes an observation of something that is phenomenologically separate from the observer and is a perceptually distinct object. By contrast, immersed experience is the kind of self-awareness one has when deeply involved in an activity that is not self-focused, such that the self is implicit and perceptually recessive. Thus, detached awareness can take a third-person form, in which case the agent adopts the third-person stance of an external observer toward his or her own activity.

Considering these two planes of analysis in greater detail, we can define the main features of the different senses of agency. In the first plane, the mental action of trying may constitute the inner, introspectively accessible aspect of feeling of agency with respect to an action. Where there is a physical action, there is a mental event of trying or willing that is commensurate with the action [6]. Events of the willing type are conscious experiences, part of the content of the stream of consciousness. In this view, we have immediate experiential knowledge only of "trying," in other words, we are introspectively aware of our actions only under descriptions of the form "I am trying to…" Thus, knowledge of displayed action is always based on *inference*. Even if the action fails (such as, in the extreme case, due to sudden paralysis), we are still aware of trying to act, as the mental event of trying also occurs in case of failed attempts.

In the second plane, a recent model hypothesized that judgment of agency, i.e., *evaluation of the agency*, is a second step in a continuous process, with the first being the non-conceptual step of *feeling the agency* [7]. Thus, if the non-conceptual feeling of agency is further processed by the cognitive system, by additionally involving conceptual capacities and belief stances, then a conceptual, interpretative judgment of being the agent is produced. What is learned on this level is to conceptually represent the effect of one's own action as exactly that. The judgment of action therefore differs from the feeling of action in the following three ways. First, judgment has an object-property structure; that is, some parts of the *conceptual self-representation* represent the system itself while others represent certain properties. Propositional I-thoughts and an explicit self-representation are included on this level. Second, conceptual agency representations are formed by inferential processes and they are influenced by other conceptual representations such as *background beliefs*. How belief formation is performed depends on how we rationalize or give plausible explanations for our experiences. Third, judgment demands the capacity to conceptually *categorize* causal forces in the interaction with the world.

But how can we compare judgment of agency with feeling of agency? A similar experimental paradigm can be used to verify both the sense of agency as a feeling subjectively experienced and the judgment of agency of the executed motor task and therefore does not discriminate between the two. The difference is instead revealed by the type of response expressed by the subject; that is, his sense of being an actor, and is thus based on immediate experience (feeling) versus the process of judging the degree of coherence between a subjective behavior and the observed action (judgment).

To explain in greater detail how the two levels of agency operate and how they communicate with each other, a synthetic review of the main research paradigms applied to the study of agency is presented in the remainder of this chapter. Specifically, we address the following points: (1) empirical research on the awareness of action, as suggested by Libet's model of action representation and action awareness; (2) the intrinsic relationship between time representation and the sense of agency; and (3) the effect of visual and auditory feedback under matching/mismatching conditions, as well as the contributions of somatosensory and body feedback to agency. It should also be noted that a second and more specific body of research has focused on the effect of feeling in agency, with reference to the illusion of subjective intentions for the agency and ownership representation, the distortion effect in the feeling of agency, i.e., when the subject is confronted with mismatching feedback, and the extended contribution of body ownership to the feeling of agency.

3.3
Empirical Paradigms of the Judgment of Agency

3.3.1
The Awareness of Action: The Contribution of Event-related Potentials

Recent research has investigated conscious awareness of the generation of movement and the relation between those conscious states and the neural processes generating movement. Specifically, Haggard and Eimer [8] evaluated the relation between neural events and the perceived time of voluntary actions or of initiating those actions.

Electrophysiological studies have examined the changes in cortical activity that precede voluntary movement and which are thought to reflect processes associated with the preparation for movement. Specifically, the relationship between intention and awareness of intentions was explored by repeating Libet's experiments (see Chapter 1). In a first series of experiments, awareness of intention was related to the *readiness potential* (RP), a brain potential associated with the specification of which of two movements to make. Generally, RP is significantly greater preceding self-paced movement than externally triggered movement, particularly when the latter is cued at unpredictable times [9]. In a second series of experiments, behavioral evidence showed an association between awareness of movement and preparation of the "motor program." In a third series, intervention in motor processing using transcranial magnetic stimulation (TMS) produced converging psychophysiological evidence that the awareness of movement was associated with brain processes concerned with the assembly and preparation of movement, rather than those concerned with execution.

However, several critical points were raised regarding Libet's experimental procedure: among others, many authors questioned whether the RP reflects specific or non-specific premotor processes. Thus, an alternative index of event-related potentials (ERPs) was proposed, the *lateralized readiness potential* (LRP), as it is a more specific index of motor preparation. Based on the data from those experiments, we can conclude the following: Firstly, both *awareness of intention* and *awareness of action* appear to occur within a narrow window of premotor processing, between the abstract prior intention to do something, and the completion of a specific program of how to do it. Awareness of intention and awareness of movement are conceptually distinct; nevertheless, they probably derive from a single processing stage in the motor pathway. Secondly, it was consistently shown that the perceived time of actions is more closely tied to movement preparation than to movement execution. Thirdly, conscious access to motor processing was restricted to the narrow window of premotor activity measured by the ERP effect (i.e., LRP). The coexistence of awareness of intention and awareness of action within a single narrow window of motor processing suggests that the binding of these two conscious representations is important. We have access to awareness of both intention and action, and the two appear to be generated by similar processing stages at comparable times in the development of action. In addition, the efferent process binding intention and awareness of action may have the dual function of bringing to consciousness the mismatch between the two, and of thus making possible a second, derived type of consciousness of the relation between my intentions and my actions. This could be a part of the sense of self.

From a neuropsychological point of view, the preparatory activity reflected in the RP is thought to arise predominantly from the supplementary motor area (SMA), lending support to evidence that this area plays a particular role in self-paced movement. The cortical source of this ERP, however, is difficult to accurately localize and the extent to which the SMA contributes to the RP has been questioned [10]. In addition, many studies have focused only on self-initiated voluntary movements and did not examine differences in the localization or timing of cortical activity generated by externally triggered movements. Studies using positron emission tomography (PET) or functional magnetic resonance imaging (fMRI) have generally found greater activation of the SMA for self-initiated movement than for externally triggered movement [11].

3.3.2
Time Perception and the Sense of Agency

Castiello and colleagues [12] designed a series of experiments to measure the *temporal dissociation* between the occurrence of an event and the subjective perception of the event itself. Interesting phenomena were revealed when the visual target to which a subject was responding was rapidly displaced.

Moreover, many studies have examined *blindsight*, adding important information on the relation between conscious/unconscious visual perception and action. Patients with lesions of the primary visual cortex appear to reach consciously for non-conscious goals. For example, patient PJG, described by Perenin and Rossetti

[13], correctly adjusted his hand movements in response to objects of varying size or orientation that were presented to his blind hemifield without being able to consciously report the presence of these objects within his visual field. More recently, Johnson et al. [14] investigated the relation between the ability to make visuomotor adjustments and the conscious experience of the adjusted movement itself. In their experiment, participants made rapid pointing movements with blocked instructions to follow the target or to move in the opposite direction. After each movement, participants were asked to reproduce the spatial path of the movement made, in this case without any time constraint. The gap between the spatial path of the original pointing movement and the spatial path of the reproduced movement was used as a measure of motor awareness. In the pointing condition, participants showed reduced and delayed motor awareness whereas in the anti-pointing condition, their corrections lacked this dissociation between performance and motor awareness. Instead, the reproduced movements indicated that participants overestimated the speed and strength of the anti-point response in the original pointing movements. Tasks such as these provide evidence that action awareness depends on what we expect to occur rather than on the physical movement of our body, supporting arguments formulated in studies of the dissociation between conscious expectancy and conditioning [15, 16]. More generally, the paradigm of Johnson et al. [14] showed that when two events appear repeatedly in succession the presentation of the first tends to modify the response to the second.

3.3.3
Visual Feedback and Awareness of Action

As previously underlined, motor performance can be distinguished from visual awareness. Indeed, *visual feedback* has often been experimentally manipulated in order to analyze the mismatch effect on the representation of the sense of agency. For example, in a series of experiments, Jeannerod [17] investigated movement awareness by instructing subjects to draw lines in the sagittal direction to a visual target using a stylus on a digital tablet. The subjects could not see their hand; only the trajectory of the stylus was visible, as a line on a computer screen, superimposed on the hand movement. A directional bias was introduced electronically, so that, in order to reach the target, the hand-held stylus had to be moved in a direction opposite to the bias. At the end of each trial, each subject was asked in which direction he thought his hand had moved by indicating the line corresponding to the estimated direction on a chart showing lines in different directions. Several important observations were made in these experiments: the subjects corrected for the bias in tracing a line that appeared visually to be directed to the target; they tended to ignore the veridical trajectory of their hand in making a conscious judgment about its direction; and they adhered to the direction seen on the screen, basing their report on visual cues and ignoring non-visual (motor or proprioceptive) signals. Thus, we can state that when biases remain small enough the visual-motor system is able to appropriately use information for producing accurate corrections, but this information is not accessed consciously. When the biases exceed a certain value, there is a strategy shift and conscious monitoring, in this case of hand movement, is used to correct for the bias. Even though the subjects in Jeannerod's experiment consciously noticed the discrepancy between what they were doing and what they saw on the screen, they experienced their movements as underestimates of their actual deviation or as being in the opposite direction to their actual movements.

The transition from automatic to conscious control can be considered firstly as a *conscious compensation strategy* [18]; secondly it may be interpreted not as the conscious detection of a discrepancy between visual and proprioceptive information but as the conscious detection of a discrepancy between the predicted and the actual visual state. In the latter case, our awareness becomes more vivid and more detailed when we are confronted with action errors too large to be automatically corrected.

A large number of our movements are prepared and executed automatically, and once started they are performed accurately and rapidly, leaving little time for top-down control. A kind of "optimization principle" is thought to intervene in ordinary movements and to operate during their execution. Optimization of execution consists of organizing and representing certain features of object-oriented movements prior to execution. This anticipatory organization can encode not only the properties of the central and peripheral motor system that optimize movement execution, but also those features of the object that are relevant to potential interactions with the agent, according to his or her intentions. In addition, processing of the object's properties must take into account the location and orientation of the object with respect to the body. Jeannerod [19] introduced the concept of *pragmatic representation* to define this mode of representing objects as goals for action. Pragmatic representations are classified as implicit functioning and are of an unconscious nature. Thus, the pragmatic level is distinct from the semantic level, which is a kind of representation formulated for identification, naming, etc., more than for action.

Indeed, an important question is whether dissociated neural pathways are associated with pragmatic and semantic representation. The classical distinction between a *dorsal visual pathway* (occipito-parietal regions) and a *ventral pathway* (occipito-temporal regions) may allow for the existence of different correlates to pragmatic/semantic representations. Clinical experience has shown that patients with lesions located in specific areas of the parietal lobe have a typical deficit in object-oriented behavior with their contralesional arm but their ability to semantically identify the object is preserved. These results support the possibility that pragmatic representation takes place in the parietal lobe, and semantic processing within the ventral stream. Another question, regarding the unconsciousness features of pragmatic/semantic representations, can be answered by considering the fact that object-oriented movements are unconscious because this is a prerequisite for accuracy. In other words, if we accept that access to conscious processing is a time-consuming affair, the necessity of accuracy does not leave enough time for the appearance of consciousness.

3.3.4
Somatosensory Information for Agency

Another study in which the correspondence between self-generated movements and their sensory effects was manipulated showed an *effect of the attenuation of sensations* due to the accuracy of sensory predictions whereas attenuation is not observed for externally generated actions [20]. For example, predictive mechanisms explain why the same tactile stimulus, such as a tickle, is felt less intensely when self-applied. This conclusion is also supported by studies in which a time delay was introduced between the motor command and the resulting tickle: the greater the time delay, the more ticklish the perception of it, probably due to a reduction in the ability to cancel the sensory feedback based on the motor command. Similarly, sensory predictions provide a mechanism to determine whether the motion of our body has been generated by us or by an external agent. When I move my arm, my predicted sensory feedback and the actual feedback match, and I therefore attribute the motion as being generated by me. However, if someone else moves my arm, my sensory predictions are discordant with the actual feedback and I attribute the movement as not being generated by me.

In a series of experiments, the authors examined whether increasing delay and trajectory perturbations increase the intensity of a tickle sensation because the stimulus no longer corresponds exactly to the efference copy produced in parallel with the motor command [20]. In that experiment, subjects held an object attached to a robot. Movements of the subject's left hand caused movement of the object by the robot, as by remote control. A robotic interface was used to introduce time delays of 100, 200, and 300 ms and trajectory rotations of 30, 60, and 90° between the movement of the participant's left hand and a tactile stimulus (tickle) on the right palm applied by the robot-held object. The subjects were then asked to rate the intensity of the tickle. As delay and rotation increased, the tickle rating increased. In other words, manipulating the correspondence between the causes and the effects of our actions deludes the motor system into treating the self as another. Thus, the attenuation of sensations, as judged by subjects' experiential accounts, is correlated with the accuracy of sensory prediction. In addition, subjects reported that they were not aware of perturbations between the movement and its consequences, which suggests that signals for sensory discrepancies are not available to our conscious awareness.

Attenuation of the perception of self-reproduced stimuli is well-documented in humans [21]. The physiological mechanisms by which this attenuation of self-produced tactile stimuli is mediated have been postulated on the basis of research on animals. Neuropsychological data demonstrated that neural responses in the somatosensory cortex are attenuated by self-generated movement. The results of fMRI demonstrate an increased activity of the primary and secondary somatosensory cortex when individuals experience an externally produced tactile stimulus on their palm relative to a self-produced tactile stimulus. What is the reason for this attenuation effect from a functional perspective? Externally produced stimuli normally carry greater biological significance than self-produced stimuli, and the actions of others are more relevant than ours. This allows, for example, unexpected stimulation to be selectively detected.

The ability to anticipate the sensory consequences of our own actions can be described using a forward model of the motor system. The forward model captures the forward or causal relationship between actions and outcomes based on an efference copy of the motor command. A computational mechanism by which the attenuation of self-produced tactile sensations might be achieved is in terms of the sensory prediction errors made by this model (see Chapter 1).

A similar paradigm was used by Wolpert and Flanagan [22] in order to explore the effect of compensation for delays in the sensorimotor system and to reduce the uncertainty in the state estimate that arises through noise inherent in both sensory and motor signals. Together with state estimation, prediction allows us to filter sensory information, attenuating unwanted information or highlighting information critical for control. Sensory prediction can be derived from the state prediction and used to delete the sensory effects of movement (re-afference) such that it is possible to cancel out the effects of sensory changes induced by self-motion, thereby enhancing more relevant sensory information (Fig. 3.2).

Recent data implicate the cerebellum and parietal cortex in sensorimotor prediction. Specifically, the cerebellum may be involved in reaching and grasping movements and it is activated before onset of the action. The cerebellum is involved in the rapid detection of errors during motor preparation and in producing error signals at an unconscious level. Differential activation of the inferior parietal cortex for tasks

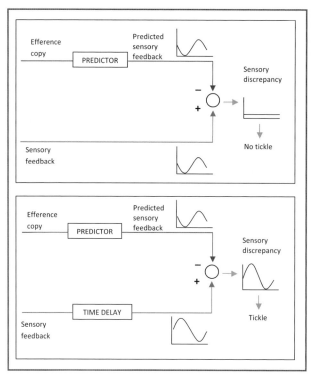

Fig. 3.2 An example of the comparator model in the presence or absence of time delay

involving action of the self *vs* action of another has been shown in neuroimaging studies. For example, greater activation of the inferior parietal cortex occurs when an external agent controls a movement than when the subject him/herself controls a movement [23]. Thus, it may be that the cerebellum participates in the rapid detection of discrepancies between actual and predicted sensory effects of movements, signaling errors below the level of awareness, while the parietal cortex is concerned with higher-level prediction, such as the maintenance of goals, the monitoring of intentions, and the distinction between self and others. This information may be available to conscious awareness [24].

3.3.5
Sense Integration

Multisensory integration appears to be intimately related to the sense of ownership of body representation [25], and intermodal matching a sufficient condition for the sense of ownership of action [26]. The relationship between *current intention*, *sensory feedback*, and *sensorimotor integration* was explored by Fink and colleagues [27], who experimentally created a conflicting and mismatch condition between vision (misleading visual feedback) and both intention and other forms of sensory feedback. Specific neural correlates were found for the mismatch condition, as the activities of the posterior parietal cortex and dorsolateral prefrontal cortex increased in response to an incongruous condition. Moreover, differential aspects of monitoring lead to differential activation within the right prefrontal cortex for an active task that emphasizes the conflict between intentions and visual and/or proprioceptive feedback, and ventrolateral prefrontal cortex activity for a comparable passive task in which the conflict is only sensory, between vision and proprioception, with no role for motor intention. It is relevant that activation of dorsolateral prefrontal cortex is associated with several functions, including complex motor selection, effort, and self-generated movement. By contrast, activation of a more ventral part of the prefrontal cortex was demonstrated in a spatial working memory task in which subjects were required to maintain but not to manipulate spatial information for brief periods of time [28].

3.3.6
Experimental Paradigms for the Feeling of Agency

The feeling of agency may be explored taking into account the subjective response to the features of agency experienced during execution of a task. Specifically, feeling of agency implies a *sense of effectiveness* in action execution that is supported by a non-reflective condition in which the individual unconsciously feels that he or she is the agent of his own action. From this perspective, the self-relation is represented in a non-conceptual, implicit manner: the idea of being the agent of an action is a non-analyzable whole, and the underlying perception of agency is not compositional and has no object-property structure. Nevertheless, this feeling may be made conscious when an individual is required to realize an explicit sense of the intended action and to report this experience. In the case of incongruity between the indicators of agency, i.e., a mismatch between proprioception, motor intention, and visual feedback, the action is experienced as strange and not fully done by me. This may result from a mismatch between efferent and afferent information.

3.3.6.1
Illusion of Intention

Wegner [29] uncovered a reconstructive mechanism of experience of intention in a study demonstrating that subjects can be led to think that they consciously intended actions or consequences of actions which they did not produce themselves. This phenomenon is said to be based on the mechanism of back referral of an intention. In a study by Wegner and Wheatley [30], subjects retrospectively attributed conscious intentions to themselves in order to explain actions that were actually performed by another person. In most of these studies, the *illusion of will* was evoked within a context in which externally produced action effects were attributed to the self. A recent contribution described the existence of differences between the confusion of intentions that may occur between the effects of self-generated and externally generated actions, and confusion about the voluntariness of our own actions [31]. Specifically, subjects may ascribe intentions to their actions, although they did not actually intend them. The subjective inability to tell the difference between a voluntary decision to resume an ongoing action and an inability to stop an ongoing action can be demonstrated by using a Go/NoGo paradigm.

In general, the introspective report of our own intentions is the product of two factors, the raw data, which is accessed via introspection, and a model, which is used to interpret the raw data. The crucial difference between introspective and objective evidence is seen in the fact that objective evidence enables the subject to refer back to the raw data [32]. According to this model, a specific, "type-C" process is included in the introspective experience but not in automatic actions. This process involves the supervisory attentional system and therefore requires accurate recollection of the presence of a decision-making process. But, since subjects are always unable to distinguish between those contexts in which they voluntary decided and those in which they failed to decide, the decision-making process is not totally intentional. This produces an important consequence regarding Libet's assumption that a veto process (introducing the possibility to control the unconsciously initiated action) can be consciously initiated: since subjects are not very accurate in observing when they have stopped a particular action, the act of vetoing cannot be consciously initiated.

An illusion of will in which we experience action outcome as initiated by ourselves although it was actually produced by another agent and an illusion of will when no other agent is involved is virtually the same: this effect is referred to as *confabulation after the fact*. But why does this experience arise? One explanation is that it might be important for people to feel that they are well-informed about their own internal processes and to know the reasons behind what they are doing. In other words, people are interested in maintaining the fiction that they have conscious will [29]. By contrast, recognizing that we are not informed about the causes of our responses makes us feel less in control of our lives, and thus less well. Reconstruction of the feeling of free choice can occur especially in situations in which we are uncertain about the degree of deliberateness of an action.

3.3.6.2
Experiencing the Disruption of Agency: Neuropsychological Evidence

In parallel with previous research on mismatching effects due to the distortion of visual feedback, a recent study by Farrer and colleagues [33] explored the feeling of the sense of agency by asking the subject to monitor the sense of control over action. That study used a visual paradigm to explore the effect of an anomalous visual feedback (angular distortion); specifically, an increased mismatch between executed and viewed action. The study assumed that the process underlying the sense of agency is not all or none but, instead, continuous and based on monitoring of the different action-related signals, which are of sensory (*visual or somatosensory*) and central (*motor command*) origin. The authors devised an experimental situation in which the visual feedback provided to the subjects about movements displayed on a computer screen was either veridical or distorted to a variable degree. The varying degree of distortion included observed movements that were completely unrelated to those actually executed. Thus, in the veridical condition, subjects were likely to feel in full control of their own movements, whereas in the maximally distorted condition they were likely to feel that they were not in control, but rather watching the movement of another agent. The results supported those of previous research: the level of activity in the main areas already shown to be activated during attribution judgments (parietal cortex and insula) varied with the amount of discordance between what the subjects did and what they saw. Specifically, a *decreasing feeling of control* due to larger degrees of distortion was associated with increased activity in the right inferior parietal lobule and, to a lesser extent, in a symmetrical zone on the left side. The graded activation of this area might therefore have been related to the increased degree of discordance between central signals arising from the motor command and visual and somatosensory signals arising from movement execution. Accordingly, the activity in the inferior parietal cortex may relate to the feeling of loss of agency associated with the discrepancy between intended actions and sensory feedback. Clinical support for this model comes from patients with lesions in this area, which are associated with delusions a patient has about, e.g., a limb, which may be perceived as an alien object or as belonging to another person. In other cases, abnormal hyperactivity in the right inferior parietal cortex has been associated with disorders of feelings of agency in psychiatric and neurological patients [34].

By contrast, in patients with decreased activity of the insula, the discordance they experience between what they do and what they see correlates with the degree of match between the different signals related to action. When the two signals are matched, activity in the insula is maximal. Farrer and Frith [23] proposed that the sense of agency is associated with a shift of attention toward representations integrating the different signals associated with the action and that this integrating process involves the insula. In their study, an explicit task was designed aimed at distinguishing self-generated from other-generated action. The results differ from those of other studies with respect to the brain areas implicated in the feeling of agency, as there was no significant contribution by the prefrontal cortex in response to intention-action match (or mismatch) [27]. This may have been due to the fact that only in the Farrer and Frith study was the subject instructed to direct his attention to the origin of the movement he saw; also, other agents were sometimes involved in the production of movement.

3.3.6.3
Embodiment or How to Represent the Self by Body Perception

Introspective experiences may be collected also in response to body perception. Bodily self-consciousness can be represented as a *non-conceptual somatic form* of knowledge, different from any other form of knowledge [35]. A recent contribution applied psychometric methods to structured introspective reports of a conscious experience of embodiment. This construct is clearly a kind of experience, but its nature is difficult to capture using traditional methods. Moreover, generally, the verbal labels that people use when describing the body enumerate the different physical parts of the body, but not the experience testifying to the fact that those parts jointly constitute the self [30]. While the objective methods of psychophysics are able to capture the occurrence of a single experience, they do not easily capture more complex experiences such as the sense of one's own body.

The heterogeneous research paradigms allow manipulation of the perceived incorporation of an external object into the representation of the body. For example, in the rubber hand illusion, a fictitious hand moving synchronously with a participant's own hand is perceived as actually being part of the participant's own body (for discussion of this concept, see also Chapter 10). This paradigm was used in a number of recent studies [36-38]; however, they simply reported the occurrence of illusion, i.e., its behavioral or neural correlates, without providing a systematic description or quantitative measurement of the changed sense of embodiment.

Longo et al. [35] investigated the structure of body and embodiment perception in a psychometric approach to introspective reports of this illusion. Both proprioceptive judgment of the location of the participants' own hand and rating of their agreement on the subjective experience of illusion were considered. Thus, the latent structure of participants' experience was explored and the complex experience of embodiment was quantified. The main structures underlying the subjective reports included four components: (1) embodiment of the rubber hand, involving subjective feelings of control and ownership of the hand; (2) loss of one's own hand, related to the disability regarding use of the hand; (3) movement, represented by perceived motion of both the subject's own hand and the rubber hand; and (4) affect, comprising aspects related to the emotional experience.

3.4
Minimal Self and Narrative Self

We have noted that the term "self-agency" is used to highlight the important distinction between the detection of agency in general and the detection of agency of oneself. While the general detection of agency (of animated objects in the environment) requires only the detection of general intentional contingencies between different entities, self-agency requires self-action, self-action perception, or at least intentional sensorimotor contingencies derived from one's sensorimotor system [7]. Thus, we can distinguish two main types of self derived from agency, the *minimal self* and the *narrative self*. The first type involves an awareness that something is occurring and the location of bodily sensations that respond to it. In this sense, "self" tends to be implicit in the particular experience. Within this category we can distinguish a sense of oneself as an agent apart from any particular action, for example, as causally effective over time, and a sense of oneself as performing a particular action at the very moment it is performed. The second type is the sense of oneself as a *distinct entity* in either the physical or the social world, and it is the core content of *autobiographical memory*. Given this distinction, the questions arise to what extent and how the different selves are linked. Although the same word (self) is used and at least superficially refers to a single entity, it is not clear whether the two types of self are mentally linked or how these forms of self consciousness developmentally arise and are experienced by the same mind.

3.4.1
Minimal Self: Self-agency as "I"

The sense of agency is the sense we experience at the time we prepare or perform a particular action [7]. Its general features are the immediate experience of the self as subject and its limits with respect to both time and that which is accessible to immediate self-consciousness. It is non-conceptual, has first person content, and is well reproduced by the use of the "I" pronoun. Even if all of the unessential features of self are stripped away, we still have an intuition that there is a basic, immediate, or primitive "something" that we are willing to call "self." Although continuity of identity over time is a major issue in the definition of personal identity, the concept of minimal self is limited to that which is accessible to immediate self-consciousness.

This type of knowledge is privileged in that it is transparent, necessarily veridical, and cannot be mistaken for something else. Moreover, it is characterized by an *immunity principle*, since when a person uses the first-person pronoun "I" he or she cannot make a mistake about the person being referred to [39]. In other words, access to the minimal self is immediate and non-observational. When I self-refer in this way, I do not go through a cognitive process in which I try to match up first-person experience with some known criterion in order to judge the experience to be my own. My access to myself in first-person experience is immediate and non-observational: it does not involve a perceptual or reflective act of consciousness. In this sense, the immediate self that is referred to here is the pre-reflective point of origin for action, experience, and thoughts.

Disruption of the immunity principle occurs in the pathological setting, in certain forms of schizophrenia. A schizophrenic patient who suffers thought insertion, for example, might claim that he is not the one who is thinking a particular thought, when in fact he is indeed the one who is thinking the thought. In general, phenomena such as delusions of control, auditory hallucinations, and thought insertion appear to involve problems with the sense of agency rather than with the sense of ownership (see also Chapter 9). In fact, there is good evidence to suggest that the sense of ownership for motor action can be explained in terms of the ecological self-awareness built into movement and perception. By contrast, experimental research on normal individuals suggests that the sense of agency for action is based on that which precedes action and translates intention into action (see also Chapter 2).

Frith's neurocognitive model [40] of the disruption of self-monitoring in schizophrenia is also a candidate for explaining immunity to error through misidentification. For example, there could be a break-down of the comparator mechanism's normal functioning. In fact, if the forward model fails or an efference copy is not properly generated, sensory feedback may still produce a sense of ownership ("I am moving") but the sense of agency will be compromised ("I am not causing the movement"), even if the actual movement matches the intended movement. Schizophrenic patients who suffer from thought insertion and delusions control could have problems with this forward, pre-action monitoring of movement, but not with motor control based on a comparison of intended movement and sensory feedback [41].

A similar model may apply to cognition and thoughts. Phenomena such as *thought insertion*, i.e., hearing voices, suggest that something is wrong with the self-monitoring mechanism. In this perspective, it is assumed not only that thinking, insofar as it is intended and self-generated, is a kind of action, but also that thinking has to match the subject's intention for it to feel self-generated, analogous to a motor action. Although such intentions are not always consciously accessible, comparator processes that match intentions to the generation of thought and to the stream of thoughts may bestow, respectively, a sense of agency and a sense of ownership for thought, as in motor action. If the mechanism that constitutes the forward aspect of this monitoring process fails, a thought occurs in the subject's own stream of consciousness but to the subject it does not seem to be self-generated or to be under his or her control. Rather, it appears to be an alien or inserted thought.

But, are there other aspects of the minimal self that are *more primitive* than those identified in the immunity principle? We have considered a self that is capable of linguistic communication, who is capable of using the first-person pronoun. If we consider that language and conceptual capacity develop in parallel, it may be that a person's immediate and pre-reflective access to self inherently involves the mediation of a conceptual framework. Is it possible to speak of non-conceptual access to the self, i.e., a more primitive self-consciousness that does not depend on the use of a first-person pronoun? This non-conceptual first-person content may consist of the self-specifying information obtained in *perceptual experience*. When I perceive objects or movements, I also gain information about myself that is pre-linguistic [42].

3.4.2
Self Ascription

I'm running, I'm young, I feel happy. What is in common to all these propositions is the subject of these sentences, the "I" reference. The presence of the first-person concept is not sufficient to make this category of I-thoughts homogeneous. How do I know I'm the person who sees the sky? I do not need to know who I am to recognize this visual experience as mine. Self-attributions of occurring mental states do not use criteria of personal identity: even if I am an amnesic I know that I see the sky. Thus, this kind of I-thoughts does not depend on any perceptual or semantic identification of the subject whereas other I-thoughts require identifying myself as the person who is described. Similarly, in order to recognize myself in the mirror, I need to identify the person that I am looking at as myself. The distinction between these two types of I-thoughts does not arise from the kind of property ascribed, but rather from the way of gaining self-knowledge. As a consequence, the same property can be self-attributed following different ways of knowing: some depend on the identification of the subject whereas others are identification-free [43]. For instance, as soon as I know the bodily property on the basis of internal information such as proprioception, I would be assured that the body that I feel is mine.

Nevertheless, we have to make a distinction between the fact that I own a certain state, mentally or bodily, and the fact that I recognize this state as mine. For instance, patients suffering from asomatognosia following a right parietal lesion deny the ownership of the limb contralateral to the brain lesion and attribute the "alien limb" to someone else or personify it. This deficit is independent from sensory deficits. Thus, we have to wonder about the nature of the cognitive conditions of the sense of ownership of one's own body.

3.4.3
Narrative Self: The Sense of Continuity

The long-term sense of agency includes a sense of oneself as an agent apart from any particular action and the sense of one's capacity for action over time, and a form of self-narrative in which past and future actions are given a general coherence through a set of goals, motives, projects, and general lines of conduct [7]. It is undeniable that we have memories, that we make plans, and that there is continuity between our past and our future.

What is the nature of this sense of a continuous self? Dennett [44] defined "self" as an abstract *center of narrative gravity*: it consists of the abstract and movable point at which the various stories that the subject tells about him/herself meet up. The notion of narrative self finds confirmation in psychology and in neuroscience. In the former, Neisser's concept of the extended and conceptual self, initially explained in terms of memory, is enhanced by considerations of the role that language and narrative play in developing our own self-concept [45]. On the other hand, recent neuropsychological approaches have led to a consensus regarding the fact that processing is for the most part distributed across various brain regions, and it cannot be said that there is a real center of experience.

Gazzaniga [15] suggested that one function of the brain's left hemisphere is to generate narratives, using the *interpreter*. In this view, the left hemisphere devises interpretations for the meanings, actions, and emotions produced by the right hemisphere. This mechanism weaves together autobiographical fact and inventive fiction to produce a personal narrative that enables a sense of a continuous self. As a whole, the self should not be false because the normal functioning of interpreter tries to make sense of what actually happens to the person. A more extended model [7] considered the self not as an abstract center but rather as an extended self, which is *decentered* and *distributed* (Fig. 3.3). This view allows for conflict, moral indecision, and self-deception in a way that would be difficult to express in terms of an abstract point of intersection. By extending the ideas of a narrative self, we are perhaps coming closer to a concept of the self that can account for the findings of the cognitive sciences and neurosciences, as well as our own experience of what it is to be a continuous self.

The long-term sense of agency contributes in great measure to a stable sense of identity and it could be extended to the self process derived from previous self-knowledge, one's own past experiences, as well as *autobiographical memories*. These memories can be recalled from either the first-person or the third-person perspective. When recalling memories from the first-person perspective, an individual sees the event associated with the memory through his or her own eyes. In contrast, when recalling memories from the third-person perspective, individuals actually see themselves in the memory, i.e., as observers watching the remembered events [46]. Conway [47] developed a comprehensive model of autobiographical memory that

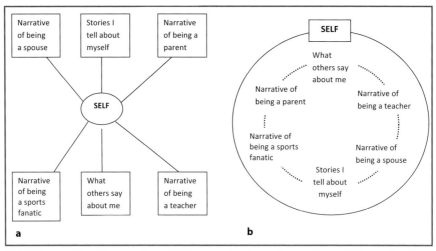

Fig. 3.3 Two models of the narrative self (**a**) as a center of narrative gravity and (**b**) according to an extended and more distributed model of the narrative self. (Modified from Gallagher [7])

emphasizes the self in memory retrieval. In this view, memory is a powerful force that acts to construct and maintain a coherent self over time. As such, memory content can be enhanced or diminished, edited or distorted, amplified or suppressed to maintain such coherence.

However, although current models emphasize the importance of self to memory retrieval, many of them do not identify or elaborate on the specific self-evaluative processes that influence the retrieval of autobiographical memories. Sutin and Robins [46] proposed a model that includes a significant *appraisal process* (Fig. 3.4), in which a fundamental step in the retrieval of autobiographical memories is to determine (appraise) whether it is relevant to the self. To carry out this appraisal, which may occur either implicitly or explicitly, individuals compare the memory to their network or self-representations, including representations of their actual, ideal, and possible selves. According to the model, memories with self relevant content are subsequently appraised for their congruence with and threat to the current self. Congruence appraisals refer to whether the self in the memory is consistent with the current self, and threat appraisals to whether the self in the memory enhances or diminishes self-esteem. Generally, individuals seek information that is consistent with their self-view. This coherence gives meaning to the self, organizes and predicts behavior, and guides social interactions. Consistent self-views promote a coherent social environment that further serves to stabilize self-views.

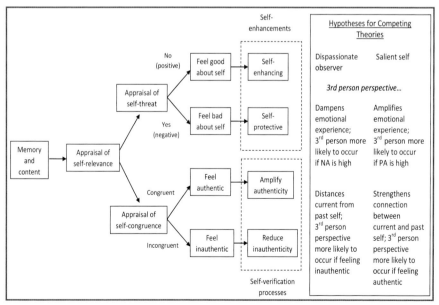

Fig. 3.4 Model of self-processes involved in the retrieval of autobiographical memories from the first-person and third-person visual perspective. *PA*, positive affect; *NA*, negative affect

References

1. Marcel AJ (2003) The sense of agency: awareness and ownership of action. In: Roessler J, Elian N (eds) Agency and self-awareness: issues in philosophy and psychology. Oxford University Press, Oxford, pp 48-93
2. Synofzik M, Vosgerau G, Newen A (2008) Beyond the comparator model: a multifactorial two-step account of agency. Conscious Cogn 17:219-239
3. Montague PR, Lohrenz T (2007) To detect and correct: norm violations and their enforcement. Neuron 56:14-18
4. Blakemore SJ, Wolpert DM, Frith CD (2002) Abnormalities in the awareness of action. Trends Cogn Sci 6:237-242
5. Slachevsky A, Pillon B, Fourneret P et al (2001) Preserved adjustment but impaired awareness in a sensory-motor conflict following prefrontal lesions. J Cognitive Neurosci 13:332-340
6. Roessler J, Elian N (2003) Agency and self-awareness: issues in philosophy and psychology. Oxford University Press, Oxford
7. Gallagher S (2000) Phylosophical conceptions of the self: implications for cognitive science. Trends Cogn Sci 4:14-21
8. Haggard P, Eimer M (1999) On the relation between brain potentials and the awareness of voluntary movements. Exp Brain Res 126:128-133
9. Cunnington R, Iansek R, Bradshaw JL, Phillips JG (1995) Movement-related potentials in Parkinson disease: presence and predictability of temporal and spatial cues. Brain 118:935-950
10. Cunnington R, Windischberger C, Deecke L, Moser E (2002) The preparation and execution of self-initiated and externally-triggered movement: a study of event-relaated fMRI. Neuroimage 15:373-385
11. Jenkins IH, Jahanshahi M, Jueptner M et al (2000) Self-initiated versus externally triggered movements. II. The effect of movement predictability on regional cerebral blood flow. Brain 123:1216-1228
12. Castiello U, Paulignan Y, Jeannerod M (1991) Temporal dissociation of motor responses and subjective awareness: a study in normal subjects. Brain 114:2639-2655
13. Perenin MT, Rossetti Y (1996) Grasping without form discrimination in a hemianopic field. Neuroreport 7:793-797
14. Johnson H, van Beers R, Haggard P (2002) Dissociations of perceptual awareness, motor awareness and motor performance during visuomotor adjustments. J Cognitive Neurosci 8:39-44
15. Perruchet P, Cleeremans A, Destrebecqz A (2006) Dissociating the effects of automatic activation and expectancy on reaction times in a simple associative learning task. J Exp Psychol Learn 32:955-965
16. Sarrazin JC, Cleeremans A, Haggard P (2008) How do we know what we are doing? Time, intention and awareness of action. Conscious Cogn 17:602-615
17. Fourneret P, Jeannerod M (1998) Limited conscious monitoring of motor performance in normal subjects. Neuropsychologia 36:1133-1140
18. Jeannerod M (2006) Motor cognition. Oxford University Press, Oxford
19. Jeannerod M (1994) The representing brain: neural correlates of motor intention and imagery. Behav Brain Sci 17:187-246
20. Blakemore SJ, Frith CD, Wolpert DM (1999) Spatio-temporal prediction modulates the perception of self-produced stimuli. J Cogn Neurosci 11:551-559
21. Collins DF, Cameron T, Gaillard DM, Prochazka A (1998) Muscular sense is attenuated when humans move. J Physiol 508:635-643

22. Wolpert DM, Flanagan JR (2001) Motor prediction. Curr Biol 11:R729-R732
23. Farrer C, Frith CD (2002) Experiencing oneself vs another person as being the cause of an action: the neural correlates of the experience of agency. Neuroimage 15:596-603
24. Blakemore SJ, Sirigu A (2003) Action prediction in the cerebellum and parietal cortex. Exp Brain Res 153:239-245
25. Armel KC, Ramachandran VS (2003) Projecting sensations to external object: Evidence from skin conductance response. P Roy Soc Lond B 270:1499-1506
26. Rochat P (1998) Self-perception and action in infancy. Exp Brain Res 123:102-109
27. Fink GR, Marshall JC, Halligan PW et al (1999) The neural consequences of conflict between intention and the senses. Brain 122:497-512
28. Jonides J, Smith EE, Koeppe RA et al (1993) Spatial working memory in humans as revealed by PET. Nature 363:623-625
29. Wegner DM (2002) The illusion of conscious will. MIT Press, Cambridge, MA
30. Wegner DM, Wheatley TP (1999) Apparent mental causation: sources of the experience of will. Am Psychol 54:480–492
31. Kühn S, Brass M (2009) Retrospective construction of the judgement of free choice. Conscious Cogn 18:12-21
32. Jack AI, Shallice T (2001) Introspective physicalism as an approach to the science of consciousness. Cognition 79:161–196
33. Farrer C, Frank N, Georgieff N et al (2003) Modulating the experience of correlates of the experience of agency. Neuroimage 15:596-603
34. Simeon D, Guralnik O, Hazlett EA et al (2000) Feeling unreal: a PET study of depersonalisation disorder. Am J Psychiat 157:1782-1788
35. Longo MR, Friederike S, Kammers MPM et al (2008) What is embodiment? A psychometric approach. Cognition 107:978-998
36. Costantini M, Haggard P (2007) The rubber hand illusion: sensitivity and reference frame for body ownership. Conscious Cogn 16:229-240
37. Farnè A, Làdavas E (2000) Dynamic size-change of hand peripersonal space following tool use. Neuroreport 11:1645-1649
38. Walton M, Spence C (2004) Cross-modal congruency and visual capture in a visual elevation-discrimination task. Exp Brain Res 154:113-120
39. Gallagher S (2008) Direct perception in the intersubjective context. Conscious Cogn 17:535-543
40. Frith CD (2005) The self in action: lessons from delusions of control. Conscious Cogn 14:752-770
41. Jeannerod M (2009) The sense of agency and its disturbances in schizophrenia: A reappraisal. Exp Brain Res 192:527-532
42. Bermúdez JL (1998) The paradox of self-consciousness. MIT Press, Cambridge, MA
43. Shoemaker S (1994) Self-knowledge and "inner sense." Philos Phenomen Res 54:249-269
44. Dennett DC (1991) Consciousness explained. Little, Brown, Boston, MA
45. Neisser U (1988) Five kinds of self-knowledge. Philos Psychol 1:35-59
46. Sutin AR, Robins RW (2008) When the "I" looks at the "Me": autobiographical memory, visual perspective, and the self. Conscious Cogn 17:1386-1397
47. Conway MA (2005) Memory and the self. J Mem Lang 53:594-628

Functional Anatomy of the Sense of Agency: Past Evidence and Future Directions

4

N. David

4.1 Introduction

Until the past decade, the sense of agency received very little attention in the field of cognitive neuroscience, despite its relevance to a variety of psychiatric and neurological syndromes associated with abnormalities in the awareness of actions [1]. Yet, compared to other areas of interest in the fields of cognitive and social cognitive neuroscience, the number of studies that have recently investigated or are currently investigating the brain basis of agency can still be considered as minor to moderate. One reason may be that the sense of agency is a topic often left to philosophers and clinicians rather than empiricists; another reason may lie in the complex, multifaceted, and often ill-defined nature of the sense of agency. Nonetheless, advancing methodologies in cognitive neuroscience and conceptual refinements of the sense of agency along with the emergence of interdisciplinary fields such as neurophilosophy have opened up new, intriguing possibilities for investigating the brain basis of agency experience.

The cognitive neuroscience approach to the sense of agency understands agency as an operationalizable construct that can be broken down into paradigms amenable to neuroscience techniques, such as functional magnetic resonance imaging (fMRI), positron emission tomography (PET), electro- or magnetoencephalograpy (EEG or MEG), and transcranial magnetic stimulation (TMS). On a general note, research strategies in cognitive neuroscience may be manifold. First, a researcher may focus on a given brain region X and investigate whether it subserves a process Y. Compared to this more *a priori* approach, a researcher may instead focus on a process Y and, in a more exploratory manner, investigate the network of brain regions associated with

---

N. David (✉)
Department of Neurophysiology and Pathophysiology, University Medical Center Hamburg-Eppendorf, Hamburg, Germany

it. In addition, a researcher may look for "loci," that is, brain regions activated during a given cognitive process (e.g., the medial temporal lobe is implicated in memory encoding), or mechanisms (e.g., how do medial temporal neurons encode new information?). However, the afore-mentioned cognitive neuroscience techniques are not equally suitable to meet the research needs posed by these different strategies; for example, fMRI is particularly–but certainly not exclusively–advantageous for addressing questions about loci, unlike EEG, which is valuable for the examination of neural mechanisms.

This chapter reviews present neuroimaging studies in the field of agency, focusing on functional anatomy. In what follows, it will become clear that the picture of the sense of agency compiled by cognitive neuroscience studies remains heterogeneous and inconclusive. The multifaceted and multilevel nature of the sense of agency may explain this heterogeneity, but it may also be due to the theoretical conceptions of the sense of agency offered by the researchers who designed these studies. It should be kept in mind that theories determine hypotheses, which determine paradigms or operationalizations, which in turn influence results. Thus, operationalizations of agency can be very diverse (Table 4.1): from manipulating the sensory feedback to a subject's movement, abolishing the sensation of self-agency and often leading to attribution of the action to another agent, to judging the onset of voluntary and involuntary movements and their sensory consequences [2].

4.2
A Functional Anatomy of the Sense of Agency: Past Evidence

Functional MRI or, in the earlier days, PET studies have associated a list of brain regions with the sense of agency, namely, the inferior parietal lobe (IPL) or posterior parietal cortex (PPC) [3-8], the cerebellum [9, 10], the posterior superior temporal sulcus (pSTS) [11, 12], the insula [3, 5], dorsolateral and ventrolateral prefrontal cortex [8, 13], as well as the supplementary motor area (SMA) [14, 15] (Fig. 4.1a).

Critical voices may consider this a rather long list of brain regions compared to the more circumscribed neural networks associated with other cognitive processes, such as episodic memory (which mainly involves medial temporal structures). However, the complex phenomenon of agency is likely to rely on lower-level sensorimotor as well as higher-level phenomenal processes [16]. To elucidate the exact functions and the specificity of each implicated region for the sense of agency, a closer inspection of the present data, especially in the context of employed tasks, seems helpful (Table 4.1) as it reveals that some brain regions have more consistently been associated with the sense of agency than others, for example, the parietal cortex and the cerebellum (Table 4.1, Fig. 4.1b). In the following, the specific contributions and possible groupings of agency-associated brain regions, displayed in Fig. 4.1, are discussed.

Table 4.1 Current agency-related neuroimaging studies and their findings

fMRI/ PET studies	Agency-related task	Activated
Blakemore et al. (1998) [31]	a. Unpredictable *vs* predictable tones, b. movement (i.e., self-generated tones) *vs* no movement conditions	a. STS, IPL b. CB, SMA, INS, DLPFC, IPL, PPC
Blakemore et al. (1998) [29]	Self- *vs* externally generated tactile stimuli	CB
Fink et al. (1999) [8]	a. Intentional hand movements under visuomotor mismatch, b. Unintentional (passive movements)	a. IPL, PPC, DLPFC b. VLPFC
Ruby and Decety, (2001) [26]	Imagined experimenter's or imagined own actions	IPL
Chaminade and Decety (2002) [7]	Leading *vs* following or observing a circle's movement	(pre-)SMA, IPL, PPC
Farrer and Frith (2002) [5]	a. Self- or b. experimenter-controlled and attributed movements	a. INS, SMA, CB b. IPL, PPC
Farrer et al. (2003) [3]	Manipulation of subjects' control of a virtual hand (a. increased control, b. decreased control)	a. INS, CB b. IPL, pre-SMA
Leube et al. (2003) [11]	Detection of temporal visuomotor mismatch during continuous hand movements	pSTS, CB
Ramnani and Miall (2004) [12]	Visually triggered button presses (third- *vs* first-person agent)	pSTS, CB
David et al. (2007) [32]	Temporally and spatially delayed visual cursor feedback *vs* real feedback to movements	IPL, SMA, VLPFC
Schnell et al. (2007) [18]	a. Monitoring and b. detection of temporal visuomotor mismatch (own *vs* computer-generated movements)	a., b. IPL a. DLPFC b. VLPFC
Farrer et al. (2008) [4]	a. Temporal delay detection task involving visuomotor mismatch, b. self-other action attribution	a., b. DLPFC, IPL a. pre-SMA
Agnew and Wise (2008) [6]	a. Active versus b. passively induced finger tapping	a. & b. SMA a. CB b. PPC
Spengler et al. (2009) [17]	Other-generated (i.e., delayed or response-incongruent) *vs* self-generated visual events	STS/ IPL

CB, cerebellum; *DLPFC*, dorsolateral prefrontal cortex; *INS*, insula; *IPL*, inferior parietal lobule; *PPC*, posterior parietal cortex; *pSTS*, posterior superior temporal sulcus; *SMA*, supplementary motor area; *VLPFC*, ventrolateral prefrontal cortex

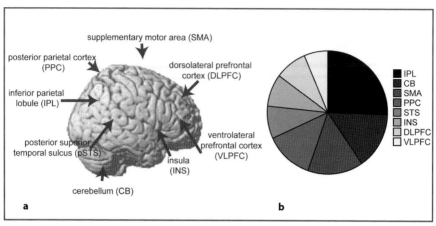

Fig. 4.1 Functional anatomy of agency. **a** *Arrows* indicate approximate anatomical locations of the brain regions implicated in the sense of agency (the right hemisphere is displayed). **b** Schematized frequencies with which these brain regions have been reported in neuroimaging studies of agency

4.2.1
The Posterior Parietal Cortex and Inferior Parietal Lobule

The parietal cortex, including its inferior (IPL) and more superior or posterior parts (PPC), is amongst the prime candidate for housing processes associated with the sense of agency. The IPL and PPC are typically activated in paradigms that employ a mismatch between predicted versus actual action outcomes or sensorimotor (especially visuomotor) incongruence [3-5,18]. These paradigms draw on an internal model originally developed for motor control [19], which in the context of agency has also been referred to as the comparator model. Essentially, this model posits that the comparison between predicted (via an efference copy) and actual sensory feedback aids the sense of agency in the following way: if the predicted and actual states agree, the sensory event is attributed to one's own agency. Incongruence between the actual and the predicted sensory feedback rather lead to the attribution of action to another agent.

In fact, as a multimodal area [20] it does not seem far-fetched that the PPC is the place in which sensory signals from many modalities, including efferent motor signals, converge and are matched. Strong support for this hypothesis also comes from patients with parietal lesions [21] and from the induction of *virtual* lesions via TMS in healthy subjects [22]. MacDonald and Paus [21] showed that, when manipulating the PPC via TMS, subjects were less likely to detect induced temporal distortions between their own finger movements and the visual feedback to these movements given by a corresponding virtual hand. The authors suggested a crucial role for the PPC in evaluating the temporal congruency of sensory and central motor signals giving rise to the sense of agency. How exactly a comparator mechanism is implemented in parietal cortices requires further research.

In addition, the role of spatial reference frames—as encoded in the parietal cortex [23]—has been discussed for the sense of agency. More specifically, it has been suggested that the actions of others are remapped and represented in allocentric, instead of egocentric, coordinates [5, 24]. In fact, a number of studies have investigated agency in relation to perspective taking [25, 26]. Given its capacity for multimodal integration, the parietal cortex may combine the different input signals from motor and sensory cortices, also taking into account different coordinate frames of signals, to produce a common, holistic spatial representation of the action.

The parietal cortex is a large area that includes action-relevant areas such as the intraparietal sulcus, somatosensory cortex (which is actually subsumed under the PPC), and the angular gyrus (as part of the IPL). To date, it is unclear whether different parietal regions, specifically, the IPL *vs* PPC, subserve different functions for the sense of agency. Both have been implicated in the detection of sensorimotor mismatch and the awareness of authorship. A more recent study has also implicated the parietal operculum in active *vs* passively induced finger movements [6]. Further research is required to answer this question.

4.2.2
The Cerebellum

It has been suggested that the cerebellum houses internal models of motor control [27]. As such, it has also been implicated in the forward and comparator models and, thus, in the prediction of the consequences of actions. Indeed, there is strong evidence for the involvement of the cerebellum in such predictions: by means of an intriguing apparatus, Blakemore et al. [28] showed increased cerebellar activation as a function of delay between predicted self-generated tactile sensation via a robotic arm and actually experienced sensations. Even stronger evidence comes from the study carried out by Synofzik and colleagues, who tested the updating of sensory predictions in patients with cerebellar lesions [33]. Although patients performed equally well detecting sensorimotor mismatch, they were impaired when they flexibly had to adapt their motor performance (i.e., update internal predictions) to implicit changes in the environment. Moreover, as another mechanism differentiating one's own actions from those of others, it has been shown that the prediction of self- but not other-generated sensory events attenuates the experience of those events (i.e., the reason why we cannot tickle ourselves but others can) [29, 30]. There is evidence that the cerebellum mediates this attenuation.

Both the parietal lobe and the cerebellum have been proposed to play a role in sensorimotor prediction [9]. How does the role of the cerebellum differ from that of the parietal cortex? In neuroimaging studies of agency (Table 4.1), the cerebellum appears to be especially activated during: (i) the experience of control over a virtual hand [3] or attribution of agency to the self [5] and, (ii) when active movements are compared to passive [6] or no [31] movements, thus possibly coding for motor efferences. By contrast, the IPL is recruited during unpredictable action outcomes [31] or the registration of sensorimotor mismatch [4, 8, 18, 32], signaling other causes (i.e., agents) for an action-related sensory event. Indeed, the IPL has explicitly been associated with the attribution of action to another agent [3–5]. It seems that the cerebellum—unlike the parietal cortex—is not necessarily associated with the actual comparison of predicted and actual signals or the detection of violations [29], nor with explicit distinctions between self- and other-generated actions. Sirigu [21] showed that patients with damage to the parietal lobule may no longer be able to differentiate their own hand movements from those of another agent (see also [22]). A similar phenomenon has not been reported for patients with damage to the cerebellum.

4.2.3
The Posterior Superior Temporal Sulcus

The role of the pSTS for the sense of agency is less clear-cut. In fact, the STS appears to be a multifunctional talent in the brain [34], implicated in social perception (e.g., of socially relevant signals such as eye gaze, hand or lip movements) [35] or other biological motion processing [36, 37], mental state inferences [38], emotions [39], and multisensory integration [40]. A few studies have also associated the pSTS with the sense of agency, more specifically, with the predictability of action consequences and violations thereof [11, 12, 17, 31]. What does the pSTS code for: the comparison of efferent and reafferent feedback signals [11]? Intentional agents other than the self [12, 41, 42]? Or simply the processing of explicit or implied biological motion [36, 43]? All of these functions certainly play into the sense of agency. Interestingly, the STS also has been associated with mirror properties, that is, shared action representation for self and other [44]. More specifically, it has been shown that neurons in the STS respond independently of who performs the action. In addition, and further elucidating the function of the STS in agency, the STS was reported to remain silent during single-cell recordings in the monkey in the absence of visual feedback to the monkey's limb movements (see [44] and compare with [45]), suggesting a predominant processing of visual and not proprioceptive or motor signals in the STS. To date, it is unclear how the pSTS could possibly subserve a mechanism, which compares predicted and actual senosry feedback or efferent and reafferent signlas thus signallig a distinction between self and other, and yet houses shared representations of self and other at the same time.

4.2.4
The Insula

Farrer and colleagues [3, 5] implicated the insula in the sense of agency (which they could not replicate in a later study) [4]. In the paradigms used by the authors, a discrepancy between the subjects' hand movements (e.g., via a joystick) and the visual feedback provided to the subjects was introduced. They found that the smaller the discrepancy, the higher the activation of the insula [3, 5], which was also associated with an increased feeling of causing the movement. Does the insula code for self-agency, as suggested by these findings? Some authors believe that the insula rather represents a correlate of body ownership (i.e., is this my hand?) instead of agency (i.e., is this my movement?) [46].

The sense of ownership can be considered closely related but distinct from the sense of agency [47]. Strong support for this alternative hypothesis on the function of the insula comes from patients with anosognosia with common damage to the insula. These patients experience their body or parts of their body as not belonging to them anymore [48, similarly 49]. Thus, the insula, as an area in which internal proprioceptive signals converge with signals from other modalities, may aid in the sense of body ownership [49].

4.2.5
The Supplementary Motor Area

The SMA shows increased activation during awareness and execution of self-generated movements [5], for example, when leading a movement versus just following/observing a movement [7] or when moving versus no movement [31]. We know that the SMA plays a crucial role not only in the execution of movements but also in their preparation and initiation [50-52]. For example, it has been shown that a pharmacologically induced temporal knockout of the SMA in monkeys has a severe impact on their ability to initiate a movement. Hence, some authors have explicitly associated the SMA with the formation of motor or action intentions and the translation of those intentions into corresponding motor commands [53]. Patients with lesions to the SMA indeed suggest a strong link between the SMA and motor intentions, often experiencing unintended actions of their own hand as if the hand had a "will of its own" [54].

4.2.6
The Prefrontal Cortex

The currently available, but limited, evidence suggests that regions in the prefrontal cortex, namely, the dorsolateral and ventrolateral prefrontal cortices, perform supervisory functions at the level of conscious monitoring of actions, leading to the conscious detection of sensorimotor mismatches or of conflict between intended/predicted and actual action outcomes [8, 18]. By contrast, these regions do not seem relevant at lower levels of agency processing, such as sensorimotor integration [55, 56]. Slachevsky and colleagues used a classical agency paradigm, which again implemented sensorimotor mismatch. They found that, unlike controls, patients with prefrontal lesions were unaware of this mismatch but nonetheless showed motor adjustments to the distorted visual feedback [55, 56]. Such findings are in line with models of consciousness in general, proposing that conscious perception is associated with increased parietofrontal activity [57].

4.3 Future Directions

The previous sections suggest the need to formulate an account of agency, one that is capable of integrating presently available evidence. In fact, recent theoretical developments followed a more holistic line, describing the sense of agency on the level of motor intentions, sensorimotor control and monitoring, as well as identification. We have learned that numerous brain regions, such as frontal, parietal, as well as cerebellar cortices, might be associated with the sense of agency. Yet, a model explaining whether or how the afore-mentioned brain regions (Fig. 4.1) interact during the sense of agency is missing. Computational models of the sense of agency, such as those developed for grasping via mirror neurons [58] and for sensorimotor control [59], may help but remain to be developed [as also discussed by 60]. Arbib and Mundhenk [41] roughly schematized a model, that integrated grasping as the action (via parietal and premotor areas), perception for the object to be grasped (via inferior temporal cortex), and encoding of the identity of the agent (according to Arbib and Mundhenk mediated by the STS, [41]). This model requires validation and further theoretical extensions.

In fact, methodological advancements allow us to test such proposed neural interactions directly. There are two possible connectivity measures of neuroimaging data [61]: (i) functional connectivity, defined by correlated time courses of activity between different brain regions, and (ii) effective connectivity containing directional assumptions (i.e., the influence one neuronal system exerts over another). Effective connectivity has been successfully applied to test interregional connectivity during grasping: Grol and colleagues [62] demonstrated increased coupling between activated parietal and frontal areas as a function of necessary online-control during grasping (e.g., bigger *vs* small objects). As mentioned above, the presently available data on the brain basis of the sense of agency are limited; we do not know enough to formulate a testable computational or connectivity model of agency. Nonetheless, potential models of dynamic interregional interactions could be proposed based on a systematic comparison of relevant fMRI studies, co-activations between different brain regions within a single neuroimaging study, and more general evidence from the monkey brain on anatomical connectivity. Regarding the latter, for example, we know that two regions implicated in the sense of agency, namely, the STS and the IPL, are reciprocally connected [63]. Connectivity analyses should guide future neuroimaging work related to the sense of agency, especially given that disconnections are thought to play a major role in disorders such as schizophrenia [64]. Of course, other neuroscience techniques such as EEG or MEG also allow for the examination of correlations or synchronous activity between distributed brain regions, and on faster time scales than fMRI [65]. In fact, the topic of neural synchrony in cortical networks recently received increased scientific attention and was investigated in relation to schizophrenia [66]. In addition, due to their excellent temporal resolution, EEG and MEG might allow the temporal-functional aspects of the sense of agency to be elucidated. These techniques have been used in studies of action observation [67, 68] but, surprisingly, not in agency research.

4.4 Conclusions

For many disciplines–be it philosophy or neuropsychology–the sense of agency is an intriguing concept. Technological advances and the emergence of interdisciplinary disciplines such as cognitive neuroscience have opened up the possibility to investigate the neurobiological underpinnings of the sense of agency, ultimately also improving our understanding of neurological-psychiatric instances of abnormalities of action awareness. Although we have started to elucidate the functional anatomy of the sense of agency, much remains to be discovered. It seems that premotor and prefrontal regions in particular code for higher-level aspects of agency as opposed to parietal-cerebellar regions; nonetheless, the exact function of these and other correlates of agency remains unclear. The limited amount of data may also contribute to the lack of computational or neural network models of the sense of agency. Thus, the investigation of functional mechanisms contributing to the sense of agency remains a widely open, interesting field of research.

References

1. Blakemore SJ, Wolpert DM, Frith CD (2002) Abnormalities in the awareness of action. Trends Cogn Sci 6:237-242
2. David N, Newen A, Vogeley K (2008) The "sense of agency" and its underlying cognitive and neural mechanisms. Conscious Cogn 17:523-534
3. Farrer C, Franck N, Georgieff N et al (2003) Modulating the experience of agency: a positron emission tomography study. Neuroimage 18:324-333
4. Farrer C, Frey SH, Van Horn JD et al (2008) The angular gyrus computes action awareness representations. Cereb Cortex 18:254-261
5. Farrer C, Frith CD (2002) Experiencing oneself vs another person as being the cause of an action: the neural correlates of the experience of agency. Neuroimage 15:596-603
6. Agnew Z, Wise RJ (2008) Separate areas for mirror responses and agency within the parietal operculum. J Neurosci 28:12268-12273
7. Chaminade T, Decety J (2002) Leader or follower? Involvement of the inferior parietal lobule in agency. Neuroreport 13:1975-1978
8. Fink GR, Marshall JC, Halligan PW et al (1999) The neural consequences of conflict between intention and the senses. Brain 122:497-512
9. Blakemore SJ, Sirigu A (2003) Action prediction in the cerebellum and in the parietal lobe. Exp Brain Res 153:239-245
10. Blakemore SJ, Oakley DA, Frith CD (2003) Delusions of alien control in the normal brain. Neuropsychologia 41:1058-1067
11. Leube DT, Knoblich G, Erb M et al (2003) The neural correlates of perceiving one's own movements. Neuroimage 20:2084-2090
12. Ramnani N, Miall RC (2004) A system in the human brain for predicting the actions of others. Nat Neurosci 7:85-90
13. Schnell K, Heekeren K, Daumann J et al (2008) Correlation of passivity symptoms and dysfunctional visuomotor action monitoring in psychosis. Brain 131:2783-2797

14. Cunnington R, Windischberger C, Robinson S, Moser E (2006) The selection of intended actions and the observation of others' actions: a time-resolved fMRI study. Neuroimage 29:1294-1302
15. Lau HC, Rogers RD, Haggard P, Passingham RE (2004) Attention to intention. Science 303:1208-1210
16. Synofzik M, Vosgerau G, Newen A (2008) Beyond the comparator model: a multifactorial two-step account of agency. Conscious Cogn 17:219-239
17. Spengler S, von Cramon DY, Brass M (2009) Was it me or was it you? How the sense of agency originates from ideomotor learning revealed by fMRI. Neuroimage 46:290-298
18. Schnell K, Heekeren K, Schnitker R et al (2007) An fMRI approach to particularize the frontoparietal network for visuomotor action monitoring: detection of incongruence between test subjects' actions and resulting perceptions. Neuroimage 34:332-341
19. Wolpert DM, Ghahramani Z, Jordan MI (1995) An internal model for sensorimotor integration. Science 269:1880-1882
20. Graziano MS (2001) A system of multimodal areas in the primate brain. Neuron 29:4-6
21. Sirigu A, Daprati E, Pradat-Diehl P et al (1999) Perception of self-generated movement following left parietal lesion. Brain 122:1867-1874
22. MacDonald PA, Paus T (2003) The role of parietal cortex in awareness of self-generated movements: a transcranial magnetic stimulation study. Cereb Cortex 13:962-967
23. Graziano MS, Gross CG (1998) Spatial maps for the control of movement. Curr Opin Neurobiol 8:195-201
24. Jeannerod M (1999) The 25th Bartlett Lecture. To act or not to act: perspectives on the representation of actions. Q J Exp Psychol A 52:1-29
25. David N, Bewernick BH, Cohen MX et al (2006) Neural representations of self versus other: visual-spatial perspective taking and agency in a virtual ball-tossing game. J Cogn Neurosci 18:898-910
26. Ruby P, Decety J (2001) Effect of subjective perspective taking during simulation of action: a PET investigation of agency. Nat Neurosci 4:546-550
27. Miall RC, Weir DJ, Wolpert DM, Stein JF (1993) Is the cerebellum a smith predictor? J Mot Behav 25:203-216
28. Blakemore SJ, Frith CD, Wolpert DM (2001) The cerebellum is involved in predicting the sensory consequences of action. Neuroreport 12:1879-1884
29. Blakemore SJ, Wolpert DM, Frith CD (1998) Central cancellation of self-produced tickle sensation. Nat Neurosci 1:635-640
30. Voss M, Ingram JN, Wolpert DM, Haggard P (2008) Mere expectation to move causes attenuation of sensory signals. PLoS ONE 3:e2866
31. Blakemore SJ, Rees G, Frith CD (1998) How do we predict the consequences of our actions? A functional imaging study. Neuropsychologia 36:521-529
32. David N, Cohen MX, Newen A et al (2007) The extrastriate cortex distinguishes between the consequences of one's own and others' behavior. Neuroimage 36:1004-1014
33. Synofzik M, Lindner A, Thier P (2008) The cerebellum updates predictions about the visual consequences of one's behavior. Curr Biol 18:814-818
34. Hein G, Knight RT (2008) Superior temporal sulcus—It's my area: or is it? J Cogn Neurosci 20:2125-2136
35. Allison T, Puce A, McCarthy G (2000) Social perception from visual cues: role of the STS region. Trends Cogn Sci 4:267-278
36. Grossman E, Donnelly M, Price R et al (2000) Brain areas involved in perception of biological motion. J Cogn Neurosci 12:711-720
37. Schultz J, Imamizu H, Kawato M, Frith CD (2004) Activation of the human superior temporal gyrus during observation of goal attribution by intentional objects. J Cogn Neurosci 16:1695-1705

38. David N, Aumann C, Santos NS et al (2008) Differential involvement of the posterior temporal cortex in mentalizing but not perspective taking. Soc Cogn Affect Neurosci 3:279-289
39. Narumoto J, Okada T, Sadato N et al (2001) Attention to emotion modulates fMRI activity in human right superior temporal sulcus. Brain Res Cogn Brain Res 12:225-231
40. Beauchamp MS, Argall BD, Bodurka J et al (2004) Unraveling multisensory integration: patchy organization within human STS multisensory cortex. Nat Neurosci 7:1190-1192
41. Arbib MA, Mundhenk TN (2005) Schizophrenia and the mirror system: an essay. Neuropsychologia 43:268-280
42. Saxe R, Xiao DK, Kovacs G et al (2004) A region of right posterior superior temporal sulcus responds to observed intentional actions. Neuropsychologia 42:1435-1446
43. Kleinhans NM, Richards T, Sterling L et al (2008) Abnormal functional connectivity in autism spectrum disorders during face processing. Brain 131:1000-1012
44. Keysers C, Perrett DI (2004) Demystifying social cognition: a Hebbian perspective. Trends Cogn Sci 8:501-507
45. Iacoboni M, Koski LM, Brass M et al (2001) Reafferent copies of imitated actions in the right superior temporal cortex. Proc Natl Acad Sci U S A 98:13995-13999
46. Tsakiris M, Hesse MD, Boy C et al (2007) Neural signatures of body ownership: a sensory network for bodily self-consciousness. Cereb Cortex 17:2235-2244
47. Gallagher S (2000) Philosophical conceptions of the self: implications for cognitive science. Trends Cogn Sci 4:14-21
48. Karnath HO, Baier B, Nagele T (2005) Awareness of the functioning of one's own limbs mediated by the insular cortex? J Neurosci 25:7134-7138
49. Berthier M, Starkstein S, Leiguarda R (1987) Behavioral effects of damage to the right insula and surrounding regions. Cortex 23:673-678
50. Cunnington R, Windischberger C, Deecke L, Moser E (2002) The preparation and execution of self-initiated and externally-triggered movement: a study of event-related fMRI. Neuroimage 15:373-385
51. Grafton ST, Mazziotta JC, Woods RP, Phelps ME (1992) Human functional anatomy of visually guided finger movements. Brain 115:565-587
52. Lee KM, Chang KH, Roh JK (1999) Subregions within the supplementary motor area activated at different stages of movement preparation and execution. Neuroimage 9:117-123
53. Haggard P (2008) Human volition: towards a neuroscience of will. Nat Rev Neurosci 9:934-946
54. Della Sala S, Marchetti C, Spinnler H (1991) Right-sided anarchic (alien) hand: a longitudinal study. Neuropsychologia 29:1113-1127
55. Slachevsky A, Pillon B, Fourneret P et al (2001) Preserved adjustment but impaired awareness in a sensory-motor conflict following prefrontal lesions. J Cogn Neurosci 13:332-340
56. Slachevsky A, Pillon B, Fourneret P et al (2003) The prefrontal cortex and conscious monitoring of action: an experimental study. Neuropsychologia 41:655-665
57. Dehaene S, Changeux JP, Naccache L et al (2006) Conscious, preconscious, and subliminal processing: a testable taxonomy. Trends Cogn Sci 10:204-211
58. Fagg AH, Arbib MA (1998) Modeling parietal-premotor interactions in primate control of grasping. Neural Netw 11:1277-1303
59. Haruno M, Wolpert DM, Kawato M (2001) Mosaic model for sensorimotor learning and control. Neural Comput 13:2201-2220
60. Schwabe L, Blanke O (2007) Cognitive neuroscience of ownership and agency. Conscious Cogn 16:661-666
61. Friston K, Büchel C (2004) Functional connectivity. In: Frackowiak RSJ (ed) Human brain function. Academic Press, San Diego
62. Grol MJ, Majdandzic J, Stephan KE et al (2007) Parieto-frontal connectivity during visually guided grasping. J Neurosci 27:11877-11887

63. Seltzer B, Pandya DN (1994) Parietal, temporal, and occipital projections to cortex of the superior temporal sulcus in the rhesus monkey: a retrograde tracer study. J Comp Neurol 343:445-463
64. Crow TJ (1998) Schizophrenia as a transcallosal misconnection syndrome. Schizophr Res 30:111-114
65. Engel AK, Fries P, Singer W (2001) Dynamic predictions: oscillations and synchrony in top-down processing. Nat Rev Neurosci 2:704-716
66. Spencer KM, Nestor PG, Perlmutter R et al (2004) Neural synchrony indexes disordered perception and cognition in schizophrenia. Proc Natl Acad Sci USA 101:17288-17293
67. van Schie HT, Koelewijn T, Jensen O et al (2008) Evidence for fast, low-level motor resonance to action observation: an MEG study. Soc Neurosci 3:213-228
68. van Schie HT, Mars RB, Coles MG, Bekkering H (2004) Modulation of activity in medial frontal and motor cortices during error observation. Nat Neurosci 7:549-554

The Monitoring of Experience and Agency in Daily Life: A Study with Italian Adolescents

M. Bassi, R.D.G. Sartori, A. Delle Fave

5.1
Agency and Its Role in Human Behavior and Experience

The sense of agency is defined as the "experience of oneself as the agent of one's own action" ([1], p. 523). It means being the one causing a specific movement or generating a certain thought in the stream of consciousness [2]. This ability implies distinguishing actions that are self-generated from those generated by others [1], thus contributing to the subjective phenomenon of self-consciousness [2-4]. Moreover, being the initiator of an action entails representation of oneself as causally responsible for the action and for its direct effects [5].

The representational content of the sense of agency is not only determined by action initiation per se, but also refers to the guidance and consequences of one's action and the causal relation between action intention, action performance, and action consequences. Agency results from the "intentional binding" of intentions, actions, and sensory feedback [6, 7]. According to Searle [8], intentions contribute substantially to action awareness; they are not separate from the action itself and include a representation of its long-term goals. Actions are continuously represented in the intention of action, through the integration of internal and external changes. In this sense, agency results from the "on-line" control of action execution. Thus, every activity, for example studying or having an interesting conversation, can be considered as a complex cognitive phenomenon characterized by action intention [6, 9, 10], knowledge [10], cues from content or the environment [10], initiation of action, awareness of movements [11], sense of activity, mental effort, sense of continuous control of action execution, and awareness of the stake of the activity as well of its long-term goals. Accordingly, the sense of agency has been studied from different perspectives and at different levels of complexity.

M. Bassi (✉)
Department of Preclinical Sciences LITA Vialba, University of Milan, Milan, Italy

Neuropsychology of the Sense of Agency. Michela Balconi (Ed.)
© Springer-Verlag Italia 2010

From the neuropsychological and the neurophysiological perspectives, tactile stimulation, proprioception (passive movements), and action (active movement) are defined as constituents of bodily awareness [12]. Several brain areas involved in the sense of agency have been detected [13-18]. As explained in other chapters of this volume, the motor system, including the ventral premotor cortex, the supplementary and pre-supplementary motor areas, the cerebellum, the dorsolateral prefrontal cortex, the posterior parietal cortex, the posterior segment of the superior temporal sulcus, and the insula are known to be involved in this complex process.

From the social-cognitive perspective, the sense of agency has been studied in terms of human development, adaptation and change [19-21]. Humans are self-organizing, proactive, self-regulating, and self-reflecting systems. In being the agent of an action, an individual intentionally influences his or her functions as well as the environment as contributions to life circumstances. Four main properties of human agency can be described within this framework: *intentionality, forethought, self-reactiveness*, and *self-reflectiveness* [20]. The first, intentionality, is the ability to form intentions that include action plans and the strategies for realizing them. An intention is a representation of a future course of action to be performed: it is not simply an expectation or prediction but a proactive commitment to bring about the action. Thus, intentions and actions are different aspects of a functional relation. The second property of human agency is forethought, which involves a temporal extension of agency that is broader than future-directed plans: people set goals and anticipate outcomes of prospective actions in order to guide and motivate their efforts. A forethought perspective provides direction, coherence, and meaning to one's life. The third property, self-reactiveness, refers to the fact that people are not only planners and forethinkers, but also self-regulators. Intentions and action plans involve the deliberate ability to make choices and organize actions, to construct appropriate courses of action, and to motivate and regulate their execution. Finally, self-reflectiveness is the ability to monitor one's own functioning: people reflect on their personal efficacy, the soundness of their thoughts and actions, and the meaning of their pursuits, making corrective adjustments if necessary.

According to Bandura [21], the core property of agency that is the most distinctly human is the metacognitive ability to reflect upon oneself and the adequacy of one's thoughts and actions. This perspective rejects the duality of human agency and social structure. Rather, it connects them in the sense that social systems are products of the human activity that organizes, guides, and regulates human behavior. Social cognitive theory thus distinguishes between individual, proxy, and collective modes of agency, which are strictly related in everyday functioning. Personal agency is exercised individually. However, in many spheres of functioning, people do not have direct control over the conditions that affect their lives. Thus, instead, they exercise socially mediated agency, or proxy agency, through their influence on others who have resources and knowledge, and act on their behalf [22-24]. Finally, many of the things people desire are attainable only by working together through interdependent effort. In the exercise of collective agency, knowledge, skills, and resources must be pooled, and actions must be undertaken in concert to shape the desired future [25].

In Bandura's *self-efficacy theory* [26], efficacy beliefs play a key role in human functioning because they affect behavior both directly and through their impact on other determinants, such as goals and aspirations, outcome expectations, and the perception of impediments and opportunities in the social environment [21]. Research findings [27, 28] confirm the crucial role of perceived self-efficacy in human activities, adaptation, and change. The core of personal beliefs is human agency, and belief in one's efficacy is a resource in personal development and change [26]. Agency operates through its impact on cognitive, motivational, affective, and decisional processes. From this perspective, individual development occurs through the interaction between the active organism and the challenges of the environment.

From a motivational and developmental perspective, Deci and Ryan's *self-determination theory* [29, 30] describes human agency as a process that refers to those motivated behaviors emanating from one's integrated self. To be agentic is to be self-determined. The distinction between autonomous and externally controlled activity is important when considering the concept of human agency. To be truly agentic means to be autonomous. As highlighted by Bandura [31], the sense of agency emerges from intentional behavior and high self-efficacy beliefs. However, people can be highly self-efficacious, believing that they can achieve whatever outcome they desire, but at the same time they can be controlled by those outcomes. In this case, they are not agentic in a true sense. By contrast, the prototype of autonomous activity, from which agency emerges as an integrated process, is the intrinsically motivated behavior that is performed out of interest and requires no separable consequence, no external or intrapsychic prods, promises, or threats [32]. Csikszentmihalyi [33] used the term "autotelic" to describe this behavior, for which the only necessary reward is the spontaneous experience of interest and enjoyment. Intrinsic motivation entails curiosity, exploration, spontaneity, and interest. Intrinsically motivated behaviors are performed when individuals are free from demands and constraints. People are primarily motivated by contexts fostering volition and self-determination [29, 30, 34-36].

Agency has also been studied in relation to cultural features. According to Markus and Kitayama [37], individuals perceive themselves as consisting of a set of attributes that enable them to be connected with or separated from others in their environment. From this perspective, agency originates either in the person or in the community, according to the culture's tendency to promote independence or interdependence. Personal agency assumes that people perceive themselves as the origin of their own behavior and are motivated to act upon opportunities that allow them to be the sole initiators of their behavior. In cultures fostering independent selves, agency is experienced as an effort to express one's internal needs, rights, and abilities, and to withstand undue social pressures. At the opposite end, in cultures fostering interdependent selves, agency is experienced as an effort to be receptive to others, to adjust to their needs, and to restrain one's own inner needs or desires [38]. Research evidence suggests that human cognition–the process that enables humans to interpret and encode information, to draw inferences, and to make judgments–is a culturally driven phenomenon [39-41]. Members of cultures promoting independence tend to build mental representations of their surroundings with regard to their components [42], focus on individual dispositions to the exclusion of other components in the environment [43-45], and attribute power and authority to the individual [46]. In contrast, people belonging to cultures promoting interdependence tend to represent their environment holistically [42, 47], make more judgments on others' behavior based on situational factors [43, 44, 48], and attribute power to the collectivity [46]. In terms of agency, cultures fostering independence assign control and stability to the individual, while those fostering interdependence attribute control to the community environment. Moreover, members of independent cultures are more motivated by contexts allowing personal agency, while in interdependent cultures members are more motivated by contexts allowing for collective agency [38].

The economist Amartya Sen [49, 50] defined agency as a constituent of eudaimonic well-being, highlighting its connection with intentionality, self-awareness, self-determination, and responsibility. The sense of agency represents the property according to which relevant and meaningful actions take into account the relation between the person, the social context, and people's needs. In the eudaimonic approach [51-53], well-being is not synonymous with pleasure, positive emotions, or needs fulfillment. Instead, it emphasizes the mobilization of resources, the development and implementation of abilities and skills, self-determined behavior, the building of social competencies and interpersonal relations, and the pursuit of aims and activities that are meaningful for the individual and society. This implies that a person can actively and voluntarily pursue activities, goals, or relations considered as important but not necessarily leading to individual benefits and pleasure. Individuals can invest personal energies and psychic and material resources into activities that are relevant for the community, sometimes sacrificing, either completely or partially, personal functioning (free time, relaxation, material goods, comforts).

5.2
Agency and Experience

In order to analyze the role of agency in human development, the relation between agency and the quality of experience in daily activities must be taken into account. The sense of agency is characterized by the monitoring of behavior in facing environmental challenges and in pursuing short- and long-term goals. In this sense, it is strictly related to the quality of experience in performing daily activities. People who actively participate in their daily activities are also agents of experiences. To successfully navigate their way through a complex world full of challenges and hazards, people have to make sound judgments about their own capabilities, anticipate the likely effects of different events and courses of action, seize opportunities available in the socio-cultural environment, and regulate their own behavior accordingly.

Research on brain development has highlighted the influence of agentic action on the functional structure of the brain [54, 55]. By regulating their motivations and activities, people produce the experiences that shape the functional neurobiological substrate of symbolic, social, and psychomotor structures, as well as other skills. Moreover, individuals have to accommodate their sense of agency with the environment and thus have to take into account experience associated with daily activities.

Exploring the relation between the quality of experience and agency can provide a bio-psycho-social understanding of human development, adaptation, and change.

5.2.1
Defining and Measuring Experience

Subjective experience is a representation of the external world based on the internal conditions of the complex human system [56, 57]. Scholars from different disciplines have emphasized the need to address experience as a unitary complex that emerges from the integration of emotional and cognitive information and which represents a subjective and dynamic representation of the world, influenced by contingent internal and external changes [58]. Experience fluctuates across different states, based on the influence of internal and external conditions. These states have been studied and defined from different disciplines.

From the neurophysiological perspective, each state of consciousness stems from the moment-by-moment integration of specific groups of neurons that interact through a rich and diverse repertoire of neural patterns. Each specific experience can therefore be considered as a well-defined configuration of neural activities [59].

From the psychological perspective, experience has been investigated in various ways. In particular, daily diaries and time-sampling procedures–which allow researchers to gather on-line information on the fluctuation of subjective states during real life–have provided new insights into the structure of experience [60]. The advantages and pitfalls of these instruments have been widely discussed [61, 62], but one of their major strengths is their effectiveness in highlighting the dynamic features of daily experience. The findings obtained through time-sampling procedures suggest that: (a) experience is idiosyncratic, in that it is related to stable individual features; (b) experience ceaselessly changes according to its contents, which are related to the contingent environmental and individual conditions; and (c) in the long term, experience contributes to broadening the behavioral repertoire available to individuals, thus promoting development and complexity at both the biological and the psychological levels [63, 64].

Interesting findings on the fluctuation of experience in daily life have been obtained through the use of the experience sampling method (ESM), a procedure developed by Csikszentmihalyi, Larson, and Prescott [65]. The ESM allows researchers to explore experience fluctuations through on-line self-reports filled out by the participants in their daily contexts. In a standard ESM session, for one week participants are provided with a booklet of forms and an electronic device sending random acoustic signals 6–8 times a day, from 8.00 am to 10.00 pm. At signal reception participants are expected to fill in a form. They are asked to describe the ongoing activities, location, and social context through open-ended questions. They are also asked to evaluate their experience using 0–12 Likert-type scales assessing the levels of cognitive, affective, and motivational variables, and of the perceived challenges and skills [66-68]. Before data analysis, forms completed more than 20 min after signal receipt are discarded, in order to avoid distortions due to retrospective recall. Answers to open-ended questions are assigned a numeric code and grouped into larger functional categories [69]. Given repeated sampling, scaled variables are usually standardized for each individual based on the weekly mean value for each variable.

ESM has been widely used across cultures, in both cross-sectional and longitudinal investigations, with clinical and non-clinical samples [70-73]. Several methodological studies have investigated its reliability and validity, as well as participants' compliance. Reliability was analyzed with test-retest split-half procedures [74-76]. As concerns validity, several studies correlated ESM data on the internal states with individual physical conditions. For example, Hoover [77] obtained high correlations between physiological indices (cardiac and motor frequency) and the ESM variables "active" and "awake". Finally, participants' compliance has been widely studied over the last two decades [75, 78-84]. The ESM procedure can be successfully used with different typologies of participants, provided that they are able to write and that researchers establish a good research alliance with them. So far, tested samples have comprised groups of people from 10 to 85 years of age [85], widely varying in their socio-demographic features [75, 78-84]. The rate of compliance shows some variations according to sample characteristics: over a one-week ESM session, Csikszentmihalyi and Larson [67] reported a signal response rate of 73% among blue-collar workers, 83% in a group of white-collar workers, and 92% among managers.

Due to repeated sampling, ESM enables researchers to detect changes in experience (namely, in the values of affective, cognitive, and motivational variables) across situations. Such changes are related to ongoing activities, social contexts, and locations. However, using the ESM it is also possible to investigate specific patterns of experience fluctuation related to subjective dimensions. In particular, several studies have shown that the quality of experience varies according to the relationship between perceived values of *environmental challenges* and of *personal skills* [87]. Specific experience fluctuation patterns related to different challenges/skills ratios have been identified through the experience fluctuation model (EFM; Fig. 5.1) [88, 89].

Depending on the challenges/skills ratio (standardized scores), in the EFM, the Cartesian plane is divided into eight areas, called channels. Four main experiences are identified, corresponding to specific channels: (a) optimal experience (channel 2), characterized by above average values of perceived challenges and skills; (b) relaxation (channel 4), associated with below-average challenges and above-average skills; (c) apathy (channel 6), characterized by below-average challenge and skill values; and (d) anxiety (channel 8), associated with above-average challenges and below-average skills. The remaining challenges/skills ratios are called transition channels [90, 91], as they are associated with intermediate experiential states: arousal (channel 1), control (channel 3), boredom (channel 5), and worry (channel 7).

Studies with various samples [79, 82, 92] confirmed that in channel 2 (optimal experience) participants report the most positive and integrated state of consciousness, and in channel 6 (apathy) a negative experience of psychic disruption and disengagement [71]. In channel 4 (relaxation), participants report positive mood and confidence but low engagement; in channel 8 (anxiety), they report high engagement but also a low level of control of the situation.

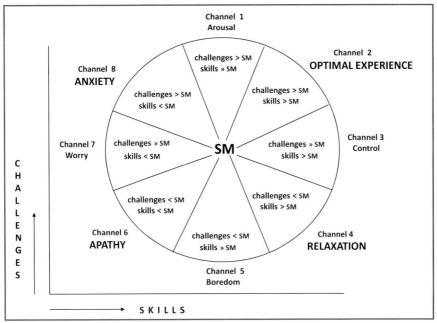

Fig. 5.1 Experience fluctuation model (EFM). *SM*, subjective mean

5.2.2
Agency in Daily Life: A Crucial Component of Optimal Experience

According to ESM findings, individuals can be aware of the moment-by-moment succession of events in consciousness and of their own intentions and action plans. From a different perspective, the neuroscience of action and the neuropsychology of schizophrenia have evidenced the existence of the "who system" [3, 93]. The *who system theory* provides a cognitive model of agency in which peripheral information from visual and proprioceptive perception of the moving body and its effects on the environment, as well as more putative central signals, are related to action initiation and the ability to plan [94, 95]. As a constituent of consciousness, agency is the sense of intending and executing actions, including the feeling of controlling one's own body movements, thoughts, events, and external environment. Agency involves a strong efferent component, because centrally generated motor commands precede voluntary movements.

Since every experience is characterized by the interplay of emotion, motivation, and cognition in response to internal and environmental events, agency, as the ability for action awareness and action monitoring, is strictly related to the quality of experience associated with every activity. A repeated sampling procedure such as the ESM allows researchers to assess this relationship, providing information on the on-line daily quality of experience and on the dimensions of agency. As previously reported, in ESM forms participants are asked to evaluate the levels of concentration, perceived goals, intention of action, sense of activity, mental effort, and situational control. Moreover, by identifying, through the EFM, different patterns of conscious experience based on the relationship between perceived challenges and skills, researchers can assess some of the key components of the sense of agency and their changes across patterns of experience fluctuation.

In the following, we specifically discuss the relationship between agency and a particular state of consciousness, optimal experience, which corresponds to channel 2 of the EFM (Fig. 5.1) and is therefore characterized by above-average challenges and skills. Optimal experience was first identified through interviews with people involved in complex and challenging tasks during their work or leisure life, such as surgery, arts, mountain climbing, or chess playing [33]. These people unanimously reported experiences of deep involvement in performing such tasks; more specifically, they described a state of consciousness characterized by deep concentration, absorption, enjoyment, control of the situation, clear-cut feedback on the course of the activity, clear goals, and intrinsic reward. Csikszentmihalyi labeled this positive and complex condition as "flow" (or optimal experience). The term "flow" synthetically expressed the feeling of fluidity and continuity in concentration and action described by most participants.

Several cross-cultural studies, conducted on samples widely differing in age, educational level, and occupation, have shown that optimal experience can occur during a wide variety of daily-life activities, such as work, study, parenting, sports, arts and crafts, social interactions, and religious practice [70, 71, 80, 82, 87]. However, regardless of the activity, the onset of optimal experience is associated with a specific condition: the ongoing task has to be challenging enough to require concentration and engagement, and to promote satisfaction in the use of personal skills. Repetitive and low-information tasks are seldom associated with flow, while its occurrence during complex activities requiring specific resources, autonomous initiative, and focused attention has been widely reported [64].

Optimal experience shapes individuals' long-term goals and competences by virtue of its dynamic structure, embedded in the perceived match between challenges and skills. This match is not stable: while first engaging in a new activity, people usually perceive challenges as much higher than their abilities, cultivation of the activity promotes the increase in related skills and the search for higher challenges, giving rise to a virtuous circle fostering the acquisition of new information. In this perspective, optimal experience promotes progressively higher competences and increasing integration of information [60]. Moreover, while optimal experience emerges from a complex integration between cognitive, emotional, and motivational components [82], it is not a peak condition. Instead, it represents a state of balance, in which all psychological components show positive values, allowing for high performance and integration of information. In the description of this experience, people do not emphasize the emotional aspect, rather the focus is the involvement in high-level external challenges, which requires active participation, and the satisfaction derived from the increase in personal abilities.

Some of the features of optimal experience are specifically related to the sense of agency:

(a) There is a merging of action and awareness; individuals are absorbed in the task at hand but at the same time report high levels of alertness and activation.
(b) Individuals are in control of the action, without the need for self-monitoring.
(c) The experience is autotelic; individuals are intrinsically motivated and are not concerned with external rewards. The primary reward consists in performing the activity itself [29, 32].
(d) The situation is characterized by clear rules and provides clear feedback on the performance.
(e) Individuals perceive clear goals in the short and in the long term.
(f) The balance between environmental challenges, perceived as stimulating and demanding, and personal skills, perceived as adequate to the challenges, promotes the effortless flow of concentration and absorption in the activity.

5.3
Empirical Evidence: A Study with Italian Adolescents

Agency refers to acts done intentionally. According to Bandura [20], an intention is "a presentation of a future course of action to be performed... a proactive commitment to bringing future actions about" (p. 6). Humans are "producers of experiences and shapers of events" ([25], p. 75). An agentic person is thus one who plans future activities, anticipates consequences, and adjusts the course of action, in order to achieve goals in different life domains. These topics are of particular interest in adolescence, a period characterized by disengagement from childhood play and introduction into the world of adult challenges [96]. Teenagers face multiple challenges and opportunities for action and engagement in their daily environment: school and learning tasks, social relations with adults and peers, and a large amount of free time—at least in post-industrial societies—which provides them with freedom from adult control and with the chance of discovering the pleasure of self-regulated experiences [97].

Several studies on adolescents focused on the sense of agency and goals [98], with the latter generally defined as cognitive representations of desired future outcomes [99]. Research identified two typologies of goals: the first and more adaptive one refers to *mastery goals*, which individuals pursue to acquire and process new information [98]. The second and less adaptive typology refers to *performance goals*, which people pursue in order to display their ability [100-104].

As previously noted, goal orientation relates to the intrinsic and extrinsic motivation of individuals for performing activities [29, 105]. Intrinsic motives are pursued for their inherent values (for example, enjoyment, interest, personal growth, social connection, and community contribution), while extrinsic motives are externally driven (for example by financial success, attractive images, fame, and popularity). Research on school motivation in adolescents found that intrinsic motives are related to more effective learning and general well-being, whereas extrinsic motives can undermine these processes [106, 107]. Another study found that mastery goals are more intrinsic, and performance goals more extrinsic [108]. Similarly, the association of learning activities with optimal experience has both short-term consequences with respect to intrinsic reward, and far-reaching implications in promoting longitudinal coherence in the amount of time devoted to studying [28, 109], in shaping individual long-term goals [60, 110], and in predicting the level of academic achievement [109, 111, 113].

Bassi and colleagues [28] investigated learning activities and the associated quality of experience of students with different levels of perceived self-efficacy. The study emphasized the long-term and day-to-day meaning students attach to learning tasks from the subjective agentic perspective, shared by perceived self-efficacy and optimal experience constructs. The authors found that low-self-efficacy participants spent significantly less time doing their homework and eschewed studying and exam preparation to the advantage of other activities. In particular, they were significantly more often involved in relaxing and low-challenging maintenance activities than high-self-efficacy students. Moreover, high-self-efficacy students mostly associated class work and homework with optimal experience, while low-self-efficacy students did not perceive a great amount of opportunities for optimal experience in learning tasks. During class work, these latter students also reported a relatively high frequency of apathy, showing cognitive, affective, and motivational disengagement from activities they perceived as unchallenging and imposed upon them.

In the domain of leisure, structured activities, such as sports, games, arts, and hobbies, merge the fun and well-being of leisure with focused attention and engagement [113]. These activities have been defined as "transitional" because they maintain aspects of childhood play, providing pleasure, self-expression, and intrinsic motivation [114], while also promoting intentional effort toward well-defined goals and competencies, typical of adult behavior [115]. Other activities, such as socializing, watching television, and listening to music, provide pleasure and fun without high demands [60]. These activities, defined as "relaxed leisure" [115], do not necessarily represent opportunities for developing specific skills. The crucial aspect distinguishing transitional from relaxed leisure activities is structure, i.e., a clear set of rules and procedures that can be associated with personal engagement, concentration, and effort toward meeting challenges and achieving goals [86].

As a whole, these findings suggest that adolescents pursue multiple goals in their daily life, some of which may be intrinsic and others extrinsic, some short-term and others long-term, related to mastery and to performance [116]. From this perspective, the study of self-regulation in terms of enhancing personal competence and gratification should also include the sense of agency in daily life, through the analysis of the related experiential variables across different states, and in particular during optimal experience. The research described in this chapter can be considered as an initial attempt to join these two theoretical perspectives. The next section will therefore explore the interplay between daily experience and agency across the main daily domains in which adolescents are involved.

5.3.1
Aims and Methods

Based on the theoretical background presented above, the study described herein was aimed at exploring quality of experience and sense of agency reported by Italian adolescents during their main daily activities. Data were collected from 261 Italian adolescents (116 males and 145 females) between 15 and 19 years of age (mean age = 17.2 years). Participants were high-school students from the metropolitan area of Rome and from Milan. The standard ESM procedure was applied, as described in Par. 5.2.1. For one week, participants were monitored during waking hours. Over the course of the study, they completed 10,173 self-reports (39 forms per participant on average). In data analysis, activities were grouped into the categories studying, structured leisure, interactions, watching TV, and maintenance. Studying comprised learning tasks, such as attending lessons (listening to the teacher and taking notes), class work (oral and written tests), other school activities (talking with friends), and homework (studying at home). Structured leisure included hobbies and sports, reading magazines and books, and thinking about various topics. Maintenance consisted of activities related to personal care (such as eating, drinking, relaxing).

For scaled items of the ESM, z-scores were obtained based on each individual's global mean for each item. Aggregated experiential values (mean z-scores) were calculated on the number of participants [117]. EFM (Fig. 5.1) was subsequently applied in data analysis.

For the purposes of this chapter, we focus on the results obtained for the four major channels of the model. In the Results paragraph, we present the percentage distribution of the adolescents' self-reports in the EFM channels. We then analyze participants' sense of agency across the channels. The sense of agency was identified through the following ESM cognitive and motivational variables: concentration, control, feeling active, wish to do the activity, and long-term goals. To provide a synthetic overview of the quality of experience in each channel, the emotional component of experience was analyzed through the variable "happy." Finally, we present the levels of agency associated with the main daily-activity domains in the four major channels. We performed t tests to highlight significant differences of the z-scores of the psychological variables from the subjective mean (corresponding to zero).

5.3.2
Results

Figure 5.2 shows the percentage distribution of the self-reports in the EFM channels. With similar percentages, participants reported optimal experience (17.48% of the answers), relaxation (17.05%), and apathy (16.11%). In smaller percentages, they reported arousal (13.01%), control (7.84%), boredom (10.56%), worry (10.09%), and anxiety (7.85%).

Figure 5.3 illustrates the quality of experience in the channels. Means and standard deviations are reported in Table 5.1. In channel 2, associated with optimal expe-

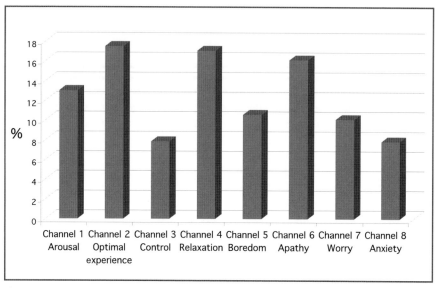

Fig. 5.2 Channel percentage distribution

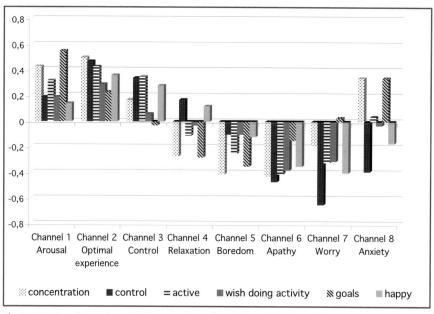

Fig. 5.3 Quality of experience in the experience fluctation model channels

Table 5.1 Quality of experience in the experience fluctation model channels (means and standard deviations)

	Channel 1 Arousal		Channel 2 Optimal experience		Channel 3 Control		Channel 4 Relaxation	
	M	Sd	M	Sd	M	Sd	M	Sd
Concentration	0.43***	0.54	0.50***	0.50	0.17**	0.74	-0.27***	0.55
Control	0.19***	0.55	0.47***	0.55	0.34***	0.67	0.17***	0.48
Happy	0.14***	0.66	0.36***	0.57	0.28***	0.69	0.12***	0.48
Active	0.32***	0.52	0.43***	0.57	0.35***	0.61	-0.11***	0.49
Wish to do the activity	0.19***	0.61	0.29***	0.49	0.06	0.65	-0.03	0.55
Goals	0.55***	0.91	0.23***	0.56	-0.03	0.68	-0.28***	0.43
N participants	238		246		207		236	

	Channel 5 Boredom		Channel 6 Apathy		Channel 7 Worry		Channel 8 Anxiety	
	M	Sd	M	Sd	M	Sd	M	Sd
Concentration	-0.41***	0.68	-0.43***	0.59	-0.19***	0.68	0.34***	0.72
Control	-0.1*	0.61	-0.47***	0.64	-0.65***	0.78	-0.39***	0.69
Happy	-0.12**	0.64	-0.35***	0.54	-0.4***	0.70	-0.17***	0.70
Active	-0.25***	0.58	-0.41***	0.58	-0.32***	0.72	0.04	0.72
Wish to do the activity	-0.1*	0.65	-0.38***	0.59	-0.31***	0.66	-0.03	0.72
Goals	-0.35***	0.40	-0.15***	0.54	0.03	0.68	0.34***	0.92
N participants	213		245		233		210	

*p<0.05; **p<0.01; ***p<0.001; M, mean; Sd, standard deviation

rience, participants reported significantly above-average values for the majority of the variables: concentration ($t=15.69$, $p<0.001$), control ($t=3.49$, $p<0.001$), active ($t=11.90$, $p<0.001$), wish to do the activity ($t=0.16$, $p<0.001$), goals ($t=4.62$, $p<0.001$), and happy ($t=9.95$, $p<0.001$). By contrast, channel 6 (apathy) was characterized by significantly below-average values of all the variables: concentration ($t=-11.55$, $p<0.001$), control ($t=-11.50$, $p<0.001$), active ($t=17.12$, $p<0.001$), wish doing the activity ($t=-10.13$, $p<0.001$), goals ($t=-3.09$, $p<0.001$), and happy ($t=-10.04$, $p<0.001$). Channel 4 (relaxation) was associated with significantly below-average values of concentration ($t=-7.52$, $p<0.001$), active ($t=-3.64$, $p<0.001$), and goals ($t=-7.03$, $p<0.001$), and significantly above-average values of happy ($t=3.68$, $p<0.001$). In channel 8 (anxiety), participants reported significantly high concentration ($t=6.92$, $p<0.001$) and goals ($t=3.80$, $p<0.001$), as well as significantly below-average values of control ($t=-8.14$, $p<0.001$) and happy ($t=-3.45$, $p<0.001$).

We then investigated the quality of experience in the four main EFM channels across the major daily activities previously described (studying, structured leisure, interactions, watching TV, and maintenance). Table 5.2 provides the data pertaining to channel 2, in which the perception of above-average challenges and skills is reported (optimal experience). Studying was associated with significantly high values of concentration ($t=7.12$, $p<0.001$), control ($t=9.50$, $p<0.001$), active ($t=8.38$, $p<0.001$), and goals ($t=6.37$, $p<0.001$). In structured leisure, adolescents reported significantly high values of concentration ($t=8.34$, $p<0.001$), control ($t=5.32$, $p<0.001$), active ($t=6.05$ $p<0.001$), wish to do the activity ($t=10.98$ $p<0.001$), goals ($t=2.20$ $p<0.03$), and happy ($t=6.02$, $p<0.001$). Participants mostly associated interactions with significantly high values of concentration ($t=3.44$, $p<0.001$), control ($t=7.39$, $p<0.001$), active ($t=8.10$, $p<0.001$), wish to do the activity ($t=8.93$, $p<0.001$), and happy ($t=8.87$, $p<0.001$).

Even if only 29.5% of the participants reported optimal experience while watching TV, most of the variables scored significantly above average: concentration ($t=4.52$, $p<0.001$), control ($t=-0.74$, $p<0.001$), active ($t=2.20$, $p<0.03$), wish to do the activity ($t=3.28$, $p<0.01$), and happy ($t=4.65$, $p<0.001$). In maintenance activities, participants reported significantly high values of control ($t=6.43$, $p<0.001$), wish to do the activity ($t=10.56$, $p<0.001$) and happy ($t=4.33$, $p<0.001$), but a significantly below average value of goals ($t=-2.26$, $p<0.03$)

Table 5.3 shows the quality of experience in channel 4, characterized by the perception of below-average challenges and above-average skills (relaxation). Studying was mostly associated with significantly below-average values of wish to do the activity and happy (respectively $t=-7.97$, $p<0.001$ and $t=-2.65$, $p<0.009$). In structured leisure, adolescents reported significantly high values of control ($t=4.77$, $p<0.001$), wish to do the activity ($t=2.49$, $p<0.014$), and happy ($t=2.54$, $p<0.013$) whereas the variable goals scored significantly below average ($t=-10.76$, $p<0.001$). During interactions, participants reported significantly below-average values of concentration ($t=-4.18$, $p<0.001$), and goals ($t=-4.43$, $p<0.001$). Happy scored significantly above average ($t=3.17$, $p<0.002$).

While watching TV, participants reported significantly above-average values of control ($t=2.95$, $p<0.01$), wish to do the activity ($t=3.77$, $p<0.001$), and happy

Table 5.2 Quality of experience in daily activities in channel 2 (optimal experience)

	Studying		Structured leisure		Interactions		Watching TV		Maintenance	
	M	Sd	M	Sd	M	Sd	M	Sd	M	Sd
Concentration	0.72***	0.59	0.56***	0.72	0.23***	0.82	0.49***	0.95	0.03	0.89
Control	0.45***	0.67	0.42***	0.83	0.43***	0.72	0.46***	0.76	0.51***	0.78
Happy	0.07	0.66	0.41***	0.73	0.61***	0.85	0.48***	0.90	0.41***	0.92
Active	0.38***	0.64	0.43***	0.75	0.47***	0.72	0.22*	0.86	0.13	1.12
Wish to do the activity	-0.08	0.77	0.56***	0.55	0.48***	0.67	0.38**	1.01	0.61***	0.57
Goals	0.55***	0.83	0.37*	1.28	0.03	0.83	0.01	1.17	-0.20*	0.62
N participants	198		113		155		77		96	

*$p<0.05$; **$p<0.01$; ***$p<0.001$; M, mean; Sd, standard deviation

Table 5.3 Quality of experience in daily activities in channel 4 (relaxation)

	Studying		Structured leisure		Interactions		Watching TV		Maintenance	
	M	Sd	M	Sd	M	Sd	M	Sd	M	Sd
Concentration	-0.03	0.80	0.14	0.86	-0.29***	0.74	-0.09	0.84	-0.59***	0.79
Control	0.05	0.83	0.34***	0.62	0.27***	0.75	0.20**	0.78	0.18***	0.68
Happy	-0.17**	0.77	0.25*	0.83	0.23**	0.79	0.22***	0.61	0.09	0.83
Active	0.03	0.80	0.06	0.81	0.10	0.75	-0.35***	0.86	-0.35***	1.06
Wish to do the activity	-0.55***	0.82	0.21*	0.72	0.05	0.83	0.23***	0.71	0.26***	0.67
Goals	-0.08	0.78	-0.38***	0.22	-0.33***	0.51	-0.4***	0.28	-0.29***	0.64
N participants	142		74		116		134		161	

*$p<0.05$; **$p<0.01$; ***$p<0.001$; M, mean; Sd, standard deviation

(t=4.08, p<0.004), whereas the variables active and goal scored significantly below average (respectively t=-4.66, p<0.001, and t=-11.75, p<0.001). In maintenance, participants reported significantly high values of control (t=3.36, p<0.001) and wish to do the activity (t=4.91, p<0.001), but significantly low values of concentration (t=-9.47, p<0.001), active (t=-4.12, p<0.001), and goals (t=-3.99, p<0.001).

Table 5.4 presents the quality of experience associated with below-average levels of challenges and skills (channel 6, apathy). Studying was mostly associated with significantly high values of goals (t=2.75, p<0.01), whereas participants reported significantly below average values of concentration (t=-4.28, p<0.001), control (t=-8.50, p<0.001), active (t=-4.84, p<0.001), and wish to do the activity (t=-17.31 p<0.001). Participants also reported significantly low values of happy (t=-8.21, p<0.001).

During structured leisure activities, adolescents reported significantly low values of active (t=-1.58 p<0.05) and goals (t=-4.09, p<0.001). Watching TV was associated with significantly low values of the variables concentration (t=-4.31, p<0.001), control (t=-3.54, p<0.001), active (t=7.59, p<0.001), and goals (t=-7.84, p<0.001). The variable happy also scored significantly below average (t=-2.83, p<0.006). Finally, during maintenance activities, participants reported significantly low values of concentration (t=-10.39, p<0.001), control (t=-6.89, p<0.001), active (t=-8.26, p<0.001), goals (t=-6.62, p<0.001), and happy (t=-5.43, p<0.001).

Table 5.5 describes the quality of experience associated with channel 8 (anxiety), in which case teenagers reported above-average challenges and below-average skills. Participants associated studying with significantly high values of concentration (t=7.91, p<0.001), active (t=2.59, p<0.05), and goals (t=5.93, p<0.001) but with below-average values of control (t=-6.10, p<0.001), wish to do the activity (t=-5.75, p<0.001), and happy (t=-4.52, p<0.001). Structured leisure was associated with significantly high values of concentration (t=3.73, p<0.001) and wish to do the activity (t=4.06, p<0.001), and with a significantly below-average value of control (t=-2.17, p<0.04).

Interactions were associated with a significantly high value of wish to do the activity (t=3.38, p<0.01), and a below-average value of control (t=-3.31, p<0.002). Only 16.8% of the participants reported anxiety while watching TV; the experience was characterized by a significantly high value of wish to do the activity (t=4.39, p<0.001) and a low value of goals (t=-2.71, p<0.01). In maintenance activities, participants reported a significantly low value of active (t=-2.24, p<0.03).

5.4
Agency and Daily Experience: A Promising Research Domain

The findings in the previous paragraph describe the quality of experience of a group of Italian adolescents and the sense of agency in their daily major domains: studying, structured leisure, interactions, watching TV, and maintenance. In line with previous studies [85], each activity presented specific experiential features that were rather stable across groups.

Table 5.4 Quality of experience in daily activities in channel 6 (apathy)

	Studying		Structured leisure		Interactions		Watching TV		Maintenance	
	M	Sd	M	Sd	M	Sd	M	Sd	M	Sd
Concentration	-0.25***	0.80	-0.19	0.97	-0.42***	0.84	-0.3***	0.76	-0.75***	0.83
Control	-0.52***	0.84	-0.24	0.97	-0.30**	0.94	-0.24***	0.75	-0.56***	0.93
Happy	-0.43***	0.73	-0.09	0.88	-0.09	0.87	-0.20**	0.78	-0.40***	0.87
Active	-0.26***	0.76	-0.20*	0.89	-0.12	0.91	-0.58***	0.83	-0.72***	1.01
Wish to do the activity	-0.88***	0.70	-0.06	0.91	-0.12	0.89	0.12	0.78	0.06	0.88
Goals	0.3**	1.06	-0.29***	0.37	-0.23**	0.88	-0.44***	0.46	-0.37***	0.49
N participants	192		48		108		119		136	

*p<0.05; **p<0.01; ***p<0.001; *M*, mean; *Sd*, standard deviation

Table 5.5 Quality of experience in daily activities in channel 8 (anxiety)

	Studying		Structured leisure		Interactions		Watching TV		Maintenance	
	M	Sd	M	Sd	M	Sd	M	Sd	M	Sd
Concentration	0.48***	0.75	0.47***	0.91	0.18	0.88	0.16	0.71	0.59	0.86
Control	-0.36***	0.73	-0.28*	0.94	-0.31**	0.89	-0.13	0.79	-1.06	0.84
Happy	-0.26***	0.71	0.07	1.07	0.04	1.02	0.05	0.93	-0.17	1.08
Active	0.14*	0.68	0.16	0.90	0.1	0.91	-0.13	1.00	-0.25*	1.12
Wish to do the activity	-0.39***	0.84	0.41***	0.73	0.31**	0.87	0.46***	0.70	-0.48	0.82
Goals	0.64***	0.95	-0.11	0.72	-0.05	0.98	-0.24*	0.45	0.35	1.58
N participants	157		53		90		44		34	

*p<0.05; **p<0.01; ***p<0.001; *M*, mean; *Sd*, standard deviation

Experience fluctuation in the four main EFM channels was investigated, focusing on some of the key variables identifying the sense of agency through cognitive dimensions (concentration, control, feeling active) and motivation (wish to do the activity and long-term goals). For the sake of completeness, the emotional component of the experience was also explored, through the variable "happy." The results showed that the perception of above-average challenges and skills (optimal experience) was associated with significantly high values of all variables related to agency. This was true of the quality of experience in general, and was confirmed in particular during major daily tasks, such as studying, structured leisure, and interactions. Watching TV and maintenance, two passive tasks characterized by a structurally low relevance to individuals' future goals, were nevertheless associated with significantly high values of most variables.

The opposite trend was detected in channel 6 (apathy). As shown in previous studies, this experience is characterized by the perception of below-average challenges and skills. The values of all the agency-related variables were lowest in apathy. By separately analyzing daily activities, however, we noted that different trends emerged. In particular, in spite of the global disruption of attention and cognitive efficiency, studying was associated with significantly high values of goals. On the contrary, this variable scored significantly below average in all the other domains, even though the quality of experience was not as globally negative as during school tasks. This can be related to the substantially intellectual features of academic learning. In activities involving a social dimension, the body, or a manual component (such as sports, hobbies, and maintenance), the disruption of experience characterized by apathy is at least partially counterbalanced by automatic routine behaviors. In these activities, which adolescents deem as less relevant for their future goals than studying (as suggested by the findings related to optimal experience), the discrepancy between expectations and intentions (assessed in terms of goals) and actual behavior (difficulty in concentrating and controlling the situation, lack of intrinsic motivation and activation) is not as dramatic as in learning tasks. This could modulate the negative effects of apathy, but further research is needed to support this hypothesis.

In channel 4 (relaxation), agency was prominently characterized by high values of control. Nonetheless, participants perceived themselves as less active, concentrated, and goal-oriented than on average. This pattern was stable across most daily activities, except in studying, in which the variable goals scored around average. In channel 8 (anxiety) participants reported significantly high values of concentration and long-term goals, both on average and with respect to a major daily activity such as studying. However, the pervasively low levels of control do not allow us to conclude that anxiety is characterized by the perception of agency.

These findings confirm several crucial concepts concerning agency [2]: awareness of action and sense of agency represent different elements of self-awareness and self-monitoring in action execution. They contribute to causing or generating an action or a certain thought in the stream of consciousness. However, they do not always need to be present together. In the complex and positive state of optimal experience, self-monitoring is absent, yet the sense of agency is strongly present and plays an important role. On the opposite end, in the case of apathy, individuals lack activation, intention and cognitive efficiency, while being aware of both their actions and their internal state.

For the variable wish to do the activity, which we can associate with the construct of intrinsic motivation, it is worth interpreting the obtained values separately. As reported in the Introduction (Par. 5.1), self-determination theory assumes that autonomous regulation is a key aspect of being agentic. However, autonomy does not only stem from the desirability of the activity in the short term (assessed by the variable "I wish to be doing the activity") but also from long-term intentions (assessed through the question on goals). The global analysis of experience fluctuation highlighted that activity desirability reaches its highest values in optimal experience and its lowest values in apathy, while in the other conditions it mostly scores on average. However, substantial differences emerged across activities as concerns goals: regardless of the quality of experience, studying was recurrently associated with the perception of above-average goals, while during watching TV and maintenance goals were never perceived as significantly above average. This finding, already noted in previous studies [60], raises the issue of long-term relevance of actions and opportunities for engagement in daily life. As suggested by Bandura, an intention is a proactive commitment to bringing about future actions [20]. However, intentions can vary according to the meaning attributed to them within the individual's and the collective value system, a dimension that should not be neglected in studies on agency.

This implication is consistent with the basic features of optimal experience. As previously stressed, this condition is not synonymous with fun. It is a rather complex and engaging state [83] whose positive features facilitate the long-term cultivation of associated activities, the shaping of individual future life goals, and the progressive unfolding of daily psychological selection [118]. Through this process of selection of environmental information, individuals preferentially replicate and cultivate subsets of opportunities for action available in the environment, thus influencing the horizontal and vertical transmission of bio-cultural information. Psychological selection results from the person's differential investment of attention and resources into daily activities [118]; therefore agency plays an important role in this process.

Humans are characterized by a constant information exchange with the environment and by the tendency towards self-organization [119]. Their biological structures are both differentiated and integrated, showing specificity of functions as well as coordination through reciprocal information exchange. The human mind emerged as an adapted learning instrument that increased the chances of survival and reproduction [60, 120-122]. Humans evolved specific psychic processes, such as awareness of the external world, awareness of one's own internal state, and higher-order consciousness [122], which is the ability to remember, make plans, and set goals through the retrieval of information acquired during interaction with the environment. This endowment promotes complexity and order in the system and enhances its adaptation to the environment. It also implies that humans never reach a stable homeostatic state; rather, they show an energetic pattern oriented toward increasing complexity [123]. They ceaselessly draw energy from the environment to preserve and enhance the inner differentiation and prevent the disruption of structures and functions.

In particular, the sense of agency promotes adaptation to the environment in terms of action awareness, continuous control of action execution, intention in action, and pursuit of long-term goals. This becomes especially evident in optimal experience, which is characterized by high complexity, effectiveness in performance, and high values of all the agency-related variables.

Conscious experience is idiosyncratic, because it results from the biological, neurophysiological, cognitive, and emotional-motivational configuration of the individual. It is also unstable: in the short term, it undergoes ceaseless transformations due to the moment-by-moment configuration of neural patterns; in the long term, it changes in relation to the progressive and continuous increase in information and complexity that characterizes human beings as organisms endowed with autopoiesis [124]. From this perspective, conscious experience, and the sense of agency as one of its core components, allows for the integration of neurobiological, psychological, and cultural information, thus fostering individual adaptation to the environment [64, 125]. A better understanding of these phenomena can help researchers identify features and trends at the psychological and cultural levels that can promote both individual complexity and cultural development.

References

1. David N, Newen A, Vogeley K (2008) The "sense of agency" and its underlying cognitive and neural mechanisms. Conscious Cogn 17:523-534
2. Gallagher S, (2000) Philosophical conceptions of the self: implications for cognitive science. Trends Cogn Sci 4:14-21
3. Georgieff N, Jeannerod M (1998) Beyond consciousness of external reality: a "Who" system for consciousness of action and self-consciousness. Conscious Cogn 7:465-477
4. Pacherie E, Jeannerod M (2004) Agency, simulation and self-determination. Mind Lang 19:113-146
5. Synofzik M, Gottfried V, Newen A (2008) Beyond the comparator model: a multifactoral two-step account of agency. Conscious Cogn 17:219-239
6. Haggard P, Clark S, Kalogeras J (2002) Voluntary actions and conscious awareness. Nat Neurosci 5:382-385
7. Haggard P, Clark S (2003) Intentional action: conscious experience and neural prediction. Conscious Cogn 12:695-707
8. Searle J (1983) Intentionality. Cambridge University Press, Cambridge
9. Aart H, Custer R, Wegner (2005) On the inference of personal authorship: enhancing experienced agency by priming effect information. Conscious Cogn 14:439-458
10. Wegner D, Sparrow B (2004) Authorship processing. In: Cazzaniga MS (ed) The new cognitive neuroscience (third ed.). MIT Press, Cambridge
11. Blakemore SJ, Frith CD, Wolpert D (1999) Spatio-temporal prediction modulates the perception of self-produced stimuli. J Cognitive Neurosci 11:551-559
12. Tsakiris M, Prabhu G, Haggard P (2006) Having a body versus moving your body: how agency structures body-ownership. Conscious Cogn 15:423-432
13. Blakemore SJ, Frith C.D, Wolpert DM (2001) The cerebellum is involved in predicting the sensory consequences of action. Neuroreport 12:1879-1884
14. Farrer C, Frith CD (2002) Experiencing oneself vs. another person as being the cause of the action: the neural correlates of the experience of agency. Neuroimage 15:596-603

15. Farrer C, Franck N, Frith CD et al (2003) Modulating the experience of agency: a positron emission tomography study. Neuroimage 18:324-333
16. Fink GR, Marshall JC, Halligan PW et al (1999) The neural consequences of conflict between intention and the senses. Brain 122:497-512
17. Jannerod M (2004) Visual and action cues contribute to the self-other distinction. Nat Neurosci 7:442-423
18. Leube DT, Knoblich G, Erb M et al (2003) The neural correlates of perceiving one's own movements. Neuroimage 20:2084-2090
19. Bandura (1986) Social foundation of thought and action: a social cognitive theory. Premice Hall, Englewood Cliffs
20. Bandura (2001) Social cognitive theory: an agentic perspective. Annu Rev Psychol 52:1-26
21. Bandura A (2006) Toward a psychology of human agency. Perspect Psychological Sci 1:164-180
22. Baltes MM (1996) The many faces of dependency in old age. Cambridge University Press, New York
23. Brandtstadter J (1992) Personal control over development: implication of self-efficacy. In: Schwarzer R (ed) Self-efficacy: through control of action. Hemisphere, Washington DC, pp 127-145
24. Ozer EM (1995) The impact of childcare responsibility and self-efficacy on the psychological health of working mother. Psychol Women Quart 19:315-336
25. Bandura A (2000) Exercise of human agency through collective efficacy. Curr Dir Psychol Sci 9:75-78.
26. Bandura A (1997) Social learning theory. Englewood Cliffs, New Jersey
27. Pastorelli C, Caprara GV, Barbaranelli C et al (2001) The structure of children's perceived self-efficacy: a cross National study. Eur J Psychol Assess 17:87-97
28. Bassi M, Steca P, Delle Fave A, Caprara GV (2007) Academic self-efficacy beliefs and quality of experience in learning. J Youth Adolescence 36:301-312
29. Deci EL, Ryan RM (1985) Intrinsic motivation and self-determination in human behavior. Plenum Press, New York
30. Deci EL, Ryan RM (2000) The "what" and "why" of goal pursuits: human needs and the self-determination of behavior. Psychol Inq 11:227-268
31. Bandura A (1989) Self-regulation of motivation and action through internal standards and goal system. In: Pervin L (ed) Goal concepts in personality and social psychology. Erlbaum Associates, Hillsdale, NJ
32. Deci EL (1975) Intrinsic motivation. Plenum Press, New York
33. Csikszentmihalyi M (2000) Beyond boredom and anxiety. Jossey-Bass, San Francisco (Original work published in 1975)
34. deCharms R (1968) Personal causation: the internal affective determinants of behavior. Academic Press, New York
35. de Charms R (1983) Personal causation. Erlbaum, New Jersey (original work published in 1968)
36. Deci EL, Ryan RM (1991) A motivational approach to self: integration in personality. In: Dienstbier R (ed) Nebraska Symposium on Motivation: Perspectives on motivation Vol. 38. University of Nebraska Press: Lincoln pp 237-288
37. Markus HR, Kitayama S (1991) Culture and the self: implications for cognition, emotion and motivation. Psychol Rev 98:224-253
38. Hernandez M, Iyengar SS (2001) What drives whom? A cultural perspective on human agency. Soc Cognition 19:269-294
39. Nisbett RE, Peng K, Choi I, Norenzayan A (2000) Culture and system of thought: holistic vs. analytic cognition. Unpublished manuscript, University of Michigan
40. Norenzayan A, Nisbett RE (2000) Culture and causal cognition. Curr Dir Psychol Sci 9:132-135

41. Peng K, Ames D, Knowles E (2002) Culture and human inference. In: Matsumoto D (ed) Handbook of cross-cultural psychology. Oxford University Press, New York
42. Masuda T, Nisbett RE (1999) Culture and attention to object versus field. Unpublished manuscript, Ann Arbor, University of Michigan
43. Miller JG (1984) Culture and the development of the everyday social explanation. J Pers Soc Psychol 46:961-978
44. Morris MW, Peng K (1994) Culture and cause: American and Chinese attributions for social and physical events. J Pers Social Psychol 67:949-971
45. Shweder RA, Bourne AJ (1984) Does the concept of the person vary cross-culturally? In: Shweder RA, LeVine RA (eds) Culture theory: essay on mind, self, and emotion. Cambridge University Press, Cambridge, pp 158-199
46. Menon T, Morris MW, Chiu CY, Hong YY (1999) Culture and construal of agency: attribution to individual versus group dispositions. J Pers Soc Psychol 76:701-717
47. Peng K, Nisbett RE (1999) Culture, dialectics, and reasoning about contradiction. Am Psychol 54:741-754
48. Shweder RA (2003) Why do men barbeque? Recipes for cultural psychology. Harvard University Press Cambridge, MA
49. Sen A (1987) On ethics and economics. Oxford University Press, Oxford
50. Sen A (1992) Inequality re-examined. Oxford University Press, Oxford
51. Deci EL, Ryan RM (2003) On assimilating identities to the self: a self-determination theory perspective on internalization and integrity within cultures. In: Leary MR, Tangney JP (eds) Handbook of self and identity. Guilford Press, New York
52. Nussbaum M (1993) Non relative virtues: an Aristotelian approach. In: Nussbaum M, Sen A (eds) The quality of life. United Nation University and WIDER, Helsinki, pp 242-269
53. Waterman AS, Schwartz SJ Conti R (2006) The implications of two conceptions of happiness (hedonic enjoyment and eudaimonia) for the understanding of intrinsic motivation. J Happiness Studies 9:41-79
54. Diamond MC (1988) Enriching heredity. Free Press, New York
55. Kolb B, Whishaw IQ (1998) Brain plastic and behavior. Ann Rev Psychol 49:43-64
56. James W (1890) The principles of psychology, vol 2. Holt, New York
57. Chalmers DJ (1995) Facing up to the problem of consciousness. J Consciousness Stud 2:200-219
58. Le Doux J (2002) Synaptic self: how our brains become who we are. Viking Penguin, New York
59. Tononi G, Edelman GM (1998) Consciousness and complexity. Science 282:1846-1851
60. Delle Fave A, Massimini F (2005) The investigation of optimal experience and apathy: developmental and psychosocial implications. Eur Psychol 10:264-274
61. Napa Scollon C, Kim-Prieto C, Diener E (2003) Experience sampling: promises and pitfalls, strenghs and weaknesses. J Happiness Stud 4:5-34
62. Gershuny J (2004) Costs and benefits of time sampling technologies. Social Indicators Research 67:247-252
63. Kunnen ES, Bosma HA (2000) Development of meaning making: a dynamic systems approach. New Ideas Psychol 18:57-82
64. Massimini F, Delle Fave A (2000) Individual development in a bio-cultural perspective. Am Psychol 55:24-33
65. Csikszentmihalyi M, Larson R, Prescott S (1977) The ecology of adolescent activity and experience. J Youth Adolescence 6:281-294
66. Csikszentmihalyi M, Larson R (1984) Being adolescent: conflict and growth in the teenage years. Basic Books, New York
67. Csikszentmihalyi M, Larson R (1987) Validity and reliability of the experience sampling method. J Nerv Mental Dis 9:526-536

68. Hektner JM, Schmidt JA, Csikszentmihalyi (2007) Experience sampling method: measuring the quality of everyday life. SAGE, Thousand Oaks, CA
69. Muhr T (1997) ATLAS/ti, User's Manual and Reference Version 4.1. Scientific Software Development, Berlin
70. Csikszentmihalyi M, Schneider B (2000) Becoming adult: how teachers prepare for the world of works. Basic Books, New York
71. Delle Fave A, Massimini F (2004) The cross-cultural investigation of optimal experience. Ric Psicol 27:79-102
72. deVries M (ed) (1992) The experience of psychopathology - Investigating mental disorders in their natural settings. Cambridge University Press, Cambridge
73. Larson RW, Richards MH (1994) Divergent realities. Basic Books, New York
74. Csikszentmihalyi M, Graef R (1980) The experience of freedom in daily life. Am J Commun Psychol 8:401-414
75. Hormuth SE (1986) The sampling experience in situ. J Pers 54:262-293
76. Larson R, Csikszentmihalyi M (1983) The experience sampling method. In: Reis H (ed) New direction for naturalistic methods in behavioural science. Jossey Bass, San Francisco
77. Hoover MC (1983) Individual differences in the relation of heart rate to self-reports. Unpublished doctoral dissertation, University of Chicago
78. Delle Fave A, Massimini F (1988) Modernization and changing context of flow in work and leisure. In: Csikszentmihalyi M, Csikszentmihalyi I (eds.) Optimal experience, psychological studies of flow in consciousness. Cambridge University Press, New York
79. Clarke S H, Haworth JT (1994) Flow experience in the daily lives of sixth-form college students. Brit J Psychol 85:511-523
80. Haworth J, Evans S (1995) Challenge, skill and positive subjective states in the daily life of a sample of YTS students. J Occup Organ Psych 68:109-121
81. Larson R, Verna S (1999) How children and adolescence spend time across the world: work, play and development opportunities. Psychol Bull 125:701-736
82. Persson D, Eklund M, Isacsson A (1999) The experience of everyday occupations and its relation to sense of coherence - a methodological study. J Occup Sci 6:13-26
83. Delle Fave A, Bassi M (2000) The quality of experience in adolescents' daily life: developmental perspectives. Soc Gen Psychol Monogr 126:347-367
84. Fianco A, Delle Fave A (2006) Donne migranti e qualità dell'esperienza soggettiva: uno studio con experience sampling method. Passaggi. Rivista Italiana di Scienze Transculturali 11:101-122
85. Larson R (1989) Beeping children and adolescents: a method for studying time use and daily experience. J Youth Adolescence 18:511-530
86. Delle Fave A, Bassi M (2003) Italian adolescents and leisure: the role of engagement and optimal experience. In: Verma S, Larson R (eds) Examining adolescent leisure time across cultures: developmental opportunities and risks. New directions in child and adolescent development. Jossey-Bass, San Francisco pp 79-93
87. Csikszentmihalyi M, Csikszentmihalyi I (eds) (1988) Optimal experience: psychological studies of flow in consciousness. Cambridge University Press, New York
88. Massimini F, Carli M (1988) The systematic assessment of daily experience. In: Csikszentmihalyi M, Csikszentmihalyi I (eds) Optimal experience: psychological studies of flow in consciousness. Cambridge University Press, New York
89. Massimini F, Csikszentmihalyi M, Carli M (1987) Optimal experience: a tool for psychiatric rehabilitation. J Nerv Ment Dis 175-179
90. Csikszentmihalyi M (1997) Intrinsic motivation and effective teaching: a flow analysis. In: Bess J (ed) Teaching well and liking it: motivating faculty to teach effectively. John Hopkins University Press, Baltimore pp 72-89
91. Delle Fave A (1996) Esperienza ottimale e fluttuazione dello stato di coscienza: risultati sperimentali. In: Massimini F, Inghilleri P, Delle Fave A (eds) La selezione psicologica umana. Cooperativa Libraria IULM, Milano, pp 541-568

92. Delle Fave A, Massimini F (1992) The ESM and the measurement of clinical change: a case of anxiety disorder. In: deVries M (ed) The experience of psychopathology: investigating mental disorders in their natural setting. Cambridge University Press, Cambridge
93. Frith CD, Blakemore S, Wolpert DM (2000) Explaining the symptoms of schizophrenia: abnormalities in the awareness of action. Brain Res Rev 31:357-363
94. Desmurget M, Grafton S (2000) Forward modeling allows feedback control for fast reaching movements. Trend Cogn Sci 4:423-431
95. Wolpert DM, Ghahramani Z (2000) Computational principles of movement neuroscience. Nat Neurosci 3:1212-1217
96. Lanz M, Iafrate R, Marta E, Rosnati R (1999) Significant others: Italian adolescents ranking compared with their parents. Psychol Rep 84:459-466
97. Olivier GGF (2000) Le ludique dans la formation sociale de l'homme: une théorie critique du loisir [Playfulness in the social formation of man: A critical theory of leisure]. Loisir et société
98. Li J (2006) Self in learning: Chinese adolescents' goals and sense of agency. Child Dev 77:482-501
99. Eccles JS, Wigfield A, Schiefele U (1998) Motivation beliefs, values and goals. Annu Rev Psychol 53:109-132
100. Ames C (1992) Classroom: goals, structures, and student motivation. J Educ Psychol 66:950-967
101. Duda JL, Nicholls JG (1992) Dimensions of achievement motivation in schoolwork and sport. J Educat Psychol 84:290-299
102. Dweck CS (1992) The study of goals in human behavior. Psychol Sci 3:165-167
103. Midgley C (1993) Motivation and middle level schools. In: Maehr ML, Pintrich PR (eds) Advantages in motivation and achievement: motivation and adolescent development, vol 8. CT:JAI, Greenwich, pp 217-274
104. Nolen SB (1989) Reasons for studying: motivational orientations and study strategies. Cognition Instruct 5:269-287
105. Ryan RM, Deci EL (2000) Self-determination theory and the facilitation of intrinsic motivation, social development and well-being. Am Psychol 55:68-78
106. Sheldon KM, Ryan RM, Deci E, Kasser T (2004) The independent effects of goal contents and motives on well-being: it's both what you pursue and why you pursue it. Pers Soc Psychol B 30:475-486
107. Vansteenkiste M, Simons J, Lens W et al (2004) Motivating learning, performance and persistence: the synergistic effects of intrinsic goal contents and autonomy-supportive contexts. J Pers Soc Psychol 87:246-260
108. Hidi S, Harackiewicz JM (2001) Motivating the academically unmotivated: a critical issue for the 21st century. Rev Educ Res 70:151-180
109. Hektner JM (1996) Exploring optimal personality development: a longitudinal study of adolescents. Unpublished doctoral dissertation, University of Chicago
110. Asakawa K, Csikszentmihalyi M (1998) The quality of experience of Asian American adolescents in activities related to future goals. J Youth Adolesc 27:141-163
111. Nakamura J (1988) Optimal experience and the use of talent. In: Csikszentmihalyi M Csikszentmihalyi I (eds) Optimal experience-psychological studies on flow in consciousness. Cambridge University Press, Cambridge, pp 319-326
112. Wong M M, Csikszentmihalyi M (1991) Affiliation motivation and daily experience: some issue on gender differences. J Pers Soc Psychol 60:154-164
113. Kleiber D, Larson R, Csikszentmihalyi M (1986) The experience of leisure in adolescence. J Leisure Res 18:169-176
114. Bassi M, Delle Fave A (2004) Adolescents and the changing context of optimal experience in time: Italy 1986-2000. J Happiness Stud 5:155-179

115. Larson R, Kleiber D (1993) Daily experience of adolescents. In: Tolan P, Cohler B (eds) Handbook of clinical research and practice with adolescents. Wiley, New York pp 125-145
116. Dowson M, McInerney DM (2003) What do student say about their motivational goals? Toward a more complex and dynamic perspective on student motivation. Contemp Educ Psychol 28:91-113
117. Larson R, Delespaul O (1992) Analysing experience sampling data: a guidebook for the perplexed. In deVries MW (ed) The experience of psychopathology: investigating mental disorders in the natural setting, Cambridge University Press, New York, pp 58-78
118. Csikszentmihalyi M, Massimini F (1985) On the psychological selection of bio-cultural information. New Ideas Psychol 3:115-138
119. Khalil EL, Boulding KE (1996) Evolution, order, and complexity. Rutledge, London
120. Changeaux JP, Chavaillon J (1995) Origins of human brain. Clarendon, Oxford
121. Nicholson N (1997) Evolutionary psychology: toward a new view of human nature and organizational society. Hum Relat 50:1050-1078
122. Edelman GM (1992) Bright air, brilliant fire. On the Matter of the Mind. Basic Books, INC trad. It: Frediani S (1993) Sulla materia della mente. Adelphi, Milano
123. Prigogine I, Stengers I (1984) Order out of chaos. Man's new dialog with nature. Bantam Books, New York
124. Maturana H, Varela F (1986) The tree of knowledge: a new look at the biological roots of human understanding. New Science Library, Boston
125. Delle Fave A (2007) Individual development and community empowerment: suggestions from studies on optimal experience. In Haworth J, Hart G (eds) Well-being: individual, community, and societal perspectives. London, Palgrave McMillan, pp 41-56

Agency and Inter-agency, Action and Joint Action: Theoretical and Neuropsychological Evidence

D. Crivelli, M. Balconi

6.1
Introduction

Agency deals with action, self-consciousness, and causality dimensions as a constitutive and pervasive aspect of human experience. The self is not a static entity but most of the time is an acting-self. To be an agent means to be in action and to encounter objects or subjects to interact with.

As underlined by van den Bos and Jeannerod: "body parts move with respect to one another and with respect to external objects as the result of intentional actions" ([1], p.178). Thus, the interactions made possible through our body become the first medium between our mind and the external world.

This chapter, after a few first philosophical remarks on agentivity and intentionality, reviews some of the most recent thoughts about joint agency. It first offers a brief definition of the concept and then goes on to present an ample discussion on intentionality and its relation to agency. It traces the path from shared collective-activity theory to joint-action theory.

Finally, it confronts the problem of agency-social relation, examining the implications of agency for social behavior and its function in interaction. The analysis includes several comments on recent research about the intersubjective and evolutionary origins of agency and its social relations.

M. Balconi (✉)
Department of Psychology, Catholic University of Milan, Milan, Italy

6.2
An Introduction to Agency

As discusses in previous chapters, agency is a classical issue in philosophical analysis but it is also of interest to cognitive and clinical neuroscience. Agency has been variably defined from different perspectives; two examples are emblematic as well as useful for the specific purposes of this chapter: the first deriving from the cognitive sciences and the second from social-cognitive theorization. Regarding the first, Gallagher singled out the concepts of sense of agency and sense of ownership in the notion of minimal self, that is, the consciousness of oneself as an immediate subject of experience, unextended in time [2]. He defined the sense of agency as the sense that I am the one who is generating an action, be it physical or mental, and causing its effects. In other words, in an ordinary situation if you execute an action or engender a thought, you know that you are the person that is acting or thinking and, simultaneously, that you intended, implemented, and thus generated that action or thought. From this point of view, the minimal self, and thus the sense of agency, plausibly hinges upon neural processes and relies on an ecologically embedded body, even if an individual need not to be aware of this in order to qualify an experience as a self-experience. In the second perspective, that of social-cognitive theory, Bandura emphasized the concept of emergent interactive agency [3]. Here, individuals are neither isolated autonomous agents nor simply mechanical conveyers of exogenous influences producing internal events, but interactive agents characterized by causal power, which contributes to their own motivation and action. In a model of triadic reciprocal causation, action, personal factors (cognitive, affective, motivational, etc.), and environmental events operate as interacting and interdependent determinants.

The cognitive perspective puts the individual at the center of the concept of agency, focusing on the link with action, the causal dimension, and personal experience. The social-cognitive perspective underlines the relation between subject and environment, highlighting the role of context and the importance of interchange between people and their specific environment for the development and emergence of agency. Bandura's definition is of merit because it focuses on the role of the social dimension besides that of action. Even if it preceded the conclusions of cognitive sciences, it is more closely in tune with the recent interests of the scientific community, i.e., the relation between agency and sociality and the different forms of collective agency or joint agency.

Having outlined the core concepts of our discussion, presenting two approaches to the agency question, we now proceed to examine theoretical advances in our understanding of the domain of collectivity.

6.3
The Beginning: Intentions and Collective Intentions

Internal models of context, of ourselves, and of our relations play a fundamental role in arguments on the sense of agency. Integrated internal representations of the outer world, of individuals as organisms, and of the relations between the two are provided, according to Newen and Vogeley [4], by self-consciousness and are based on actual experiences, perceptions, and memories. These supply reflected and adapted responses to the needs of our environment, in recognition of our bonds to it, and thus allow us to orient ourselves in the world. Both exogenous and endogenous information is elaborated during agency-related processes, with goals, intentions, efferent motor commands, reafferent sensory signals, and afferent contextual information all contributing both to our ability to act and to our sense of agency. The body tends to become integrated with intentional action, shaping an embodied self-conscious experience of action not marked by the potential unconsciousness of sensory events: "[o]riginally, my body is experienced, not as an object, but as a field of activity and affectivity, as a potentiality of mobility and volition, as an "I do" and "I can" [5]. The body image merges with the intentional structure to generate a unified agent capable of perceiving and influencing the individual and his or her environment, thus forming a complex element comprising object as well as subject and primarily involved in the construction of actions as well as interactions.

Intentionality, even if it does not by itself generate or account for agency and related processes, has been historically associated with them. The first philosophical remarks exploring the construct of agency had their roots in theorizations about individual intentions, due to the latter's strong link with action. Similarly, recent developments towards interactive perspectives of agency originated in theorizations on collective or shared intentionality.

6.3.1
From I to We

Agency is the experience of being the source and the cause of our actions—we intend, we perceive, we evaluate the context, and we act, monitoring the effect of our actions: but what about our interactions with others? When our activity becomes complex and involves more coordinated agents, how do we achieve our shared goals? What happens at the level of intentions and agency experience?

After having theorized the concepts of prior intention and intention-in-action, Searle focused on multi-agent behavior and collective intention, expressed in the form "*we* intend to do *j*" or "*we* are doing *j*" [6]. He noted that, when two or more agents act together carrying out collective intentional behavior, the simple sum of "I intend" and "I act" cannot account for "we intend" and "we act" phenomena, and that both of the "I" sentences are rather primitive. Collective intentions and behavior are, instead, specific and original in their nature. Moreover, Searle underlined that it is not possible to find any bodily movements that are movements of the various inter-agents, concluding that what makes multi-agent behavior special must be a feature of a mental component, i.e., *intentionality*. Reductionist perspectives in the discussion about collective intentions–reducing them to many pooled individual intentions plus beliefs and mutual beliefs about the intentions of other members of a group–were criticized by Searle because they do not necessarily take into account the presence of cooperation, considered as a basic notion that is implied by genuine collective intentionality.

Thus, primitive *I-intentions* may be flanked by primitive *we-intention*s. This, however, raises several questions about the structure of we-intentions and their relation to agency. The structure of we-intentions, according to Searle [6], is composed only of a single, complex intention-in-action, such as the I-intention, except that it is not of the classic type but rather an *achieve-collective-B-by-means-of-singular-A* type of intention-in-action. An example may better explain this concept. Imagine that Marc and George are cooperating in order to create a house of cards, Marc placing the "walls" and George placing the "ceilings." In this case, they have a we-intention whose content is that placing the "walls" (from Marc's point of view) or placing the "ceilings" (from George's point of view) causes the achievement of the creation of the house of cards (the collective B): the shared goal is achieved by means of each related and coordinated single action.

Searle proposed that a prerequisite of the ability to manage collective intentions and engage in collective activity is the ability to recognize the other as similar to us and as an actual or potential agent in cooperative activities, a pre-intentional ability in which benefit is achieved by acting together but does not originate from it. Collective intentionality therefore presupposes a type of communal awareness comprising reciprocal recognition of the other as an actual or potential collective agent; a kind of stance and not merely a belief. As Searle put it, "just as my stance toward others is that of their being conscious agents, without my needing or having a special belief that they are conscious" [6].Collective intentions such as "*we* intend that *we* perform j," exist in the mind of each individual agent acting in a group, i.e., involved in an interaction, even if what is primarily being referred to is a collective entity instead of an individual one and even if at that moment the co-agents are not present.

6.3.2
We in Action

Focusing specifically on the acting domain, Bratman addressed the problem of shared cooperative activities, identifying their features, the role of shared mental states, and the contribution of those states to action [8]. Bratman stated that cooperation involves and originates from intentional agents who recognize and treat others as such, respecting their intentional agency, i.e., their capability to cause or generate an action. He also underlined that cooperation implies that each agent has to accept that, in part, he or she acts because of the intention of others to act and their related subplans. These conditions favor a person's participation as an intentional agent [7].

Bratman pointed out three features characteristic of cooperative activity: *mutual responsiveness*, *commitment to the joint activity*, and *commitment to mutual support*. The first accounts for the attempts of each involved agent to be responsive to the intentions and actions of others, knowing that the others are attempting to do the same. The second means that each participant, even if for different reasons, has an appropriate commitment to joint activity, pursued through mutual responsiveness. The third relates to the commitment to support the others' efforts to play their roles in the joint activity, in order to successfully achieve the desired goal.

In shared cooperative activities, each agent intends that the group performs the joint activity in accordance with specific intentions and subplans that mesh; consequently, giving rise to a complex web of intentions. Bratman, however, strongly underlined the role of action and mutual responsiveness in action. It is exactly the joint nature of the activity that identifies an action as a cooperative one and allows the participating agents to feel as such, to experience actual interaction, and perhaps to develop an individual, or even joint, sense of agency.

6.4
Doing Things Together: Joint Action and the Sense of Agency

A discussion of our causality-in-action and our agentive stance raises notions such as intersubjectivity, in this case emphasizing its link to actual behavior and taking into account complex and collective activities. In the scientific community, increasing attention is being devoted to the inter-action issue, especially in the social neuroscience domain. This field aims at exploring the biological basis of social perception and cognition, how social behavior and context can influence short-term and long-term brain functioning, and how brain function fosters and creates social behavior and actively processes social context.

The integration of theories, hypotheses, and methods borrowed from the neurosciences and the behavioral and social sciences provides the foundation for experimental testing and increases the comprehensiveness, relevance, and impact of its evidence and of the resulting conclusions. This, in turn, allows inferences to be drawn regarding the neural correlates of social functions and the nature of active information processing. Interaction can be explored in many different ways and from different points of view—the goal being to deepen our knowledge of its implications and of the elements that make interaction, fundamental to our development and daily life, possible [8].

Social neuroscience, in particular, is aimed at better comprehending the functional and biological mechanisms that support people's ability to interact with others [9]. Besides the ability to infer people's intentions, emotions, and attitudes based on their behavior in performing particular actions, this experience activates in the individual specific representations he or she then uses to perform the observed action in his or her own action system. Different authors have hypothesized a more immediate approach to social understanding and social interaction, based on this close link between perception and action [10, 11]. Meltzoff and Moore [12] suggested the existence of an innate perception-production coupling mechanism in human beings, in which intending, simulating, observing, and performing an action have functional equivalence based on the activation of shared representation [13]. Indeed, recent empirical evidence has shown an overlap among brain regions functionally related to those different processes [14]. Moreover, action representations plausibly have a considerable value in enabling interpretation of those actions as well as the intentions of social interlocutors [15, 16].

In psycho-social theorizations, the joint-action construct involves a shift of attention away from a stimulus-receiving monad to an enlightened, open, pro-active individual, an interactive dyad. Joint actions are described as "social interactions wherein two or more individuals coordinate their actions in space and time to bring about a change in the environment" ([9], p. 100). They imply the sharing of action representations and the integrative coordination of participants' actions to achieve common goals [17, 18]. This, in turn, makes use of several sub-functions and mechanisms, in particular:

(a) joint attention, which creates a perceptual common ground in which perceptual input is shared and attention is focused on the same event;
(b) action understanding [19] and prediction [20], carried out by the activation of motor representations related to the observed actions in the observer's system [14, 21, 22] and making use of the direct link between perception and execution;
(c) creation of a task-sharing representation, which enable us to predict the actions of others based on contextual events, even without action observation [23];
(d) with the other participants, spatial and temporal action coordination in action planning to implement anticipatory action control;
(e) control of agency uncertainty, distinguishing the effects of one's own actions and those of others [24, 25].

According to Pacherie and Dokic [17], it is possible to distinguish between two forms of joint action, a *thin form* and a *thick form*, depending on the depth, awareness, and comprehension of the interaction. The thin form is defined as a superficial modality common to animals and humans; it implies cooperation without representation because it is not necessary that subject A represents his/her action as cooperative, nor the intentions of others (of B, C, ...), nor the relation between these and his/her own intentions. Thus, the thin form of joint action needs what Pacherie and Dokic define as visual and motor understanding skills, that is, the ability to feel and recognize a biological movement as such, to perceive a goal-directed act, and to acknowledge an action plan as not merely an ensemble of movements. A wolf-pack hunt and assembly-line work are two examples of this simpler form of joint action.

The thick form of joint interaction is species-specific, observable only in human groups, and requires that participants integrate the intentions system of others into his/her own, thereby reaching a complete level of cooperation, i.e., an intentional and actual cooperation that places the others in the role of active agents. This evolved form requires, besides visual and motor understanding, agentive and meta-representational understanding. In the latter, a pro-active role is attributed to the inter-agent, in recognition of the complex relationship between an agent with a goal, the instrumental means used, and the effects produced. This form of understanding is obtained by moving from intentionality-in-the-world to agent-intentionality, from goals as relational structures in the world to goals as intentional relations between the agent and the world [17]. Accordingly, the thick form of joint action, similar to Bratman's *shared cooperative activity* [7], could be considered as a more elaborate definition of the folk construct of social interaction. A worthy example of this form could be ballet, as proposed by Searle [6].

Based on its strong link with perception and action, its representational implementation, sensitivity to the actions of others, and integrative nature, the *mirror neuron system* (MNS) has been suggested to play a role in joint-action tasks [13, 18, 26]. However, as Pacherie and Dokic [17] also pointed out, social interactive behavior entails a complex representational multilevel structure, whose processing cannot be carried out solely by the MNS although it surely has a role in processes related to joint action. As noted, it may be that the MNS manages joint-action control, implementing adaptive correction mechanisms as a function of common goals and the predicted effects of others' actions, in addition to control of one's own actions. In this scenario, the MNS would foster an understanding of the actions of others. But the subject-centered nature of mirror neurons, their agent-neutrality (parameters relative to the agent are not, in fact, explicitly processed), and their elaboration of motor acts and related effects rather than intentions are such that even if they enable efficient and effective management of joint-action execution and sequence representation, they are unable to handle the joint-action process as a whole, especially with respect to coping with and integrating intentions.

Recent empirical studies have investigated the neuropsychological correlates of interaction but have reached partially contrasting conclusions: the MNS seems to be involved in joint-action tasks and is thus essential. Nevertheless, there is some evidence showing the involvement of other, neighboring cortical areas; thus, it is not clear whether the MNS alone is sufficient in the process of interaction [27-29].

6.5
Over the Self-other Differentiation: Circular Interactions and Joint Agency

As well pointed out by Semin and Cacioppo [30], social interactions, and social-cognition function, are complex *in fieri* phenomena and therefore are not limited to reception, reproduction, or representation. Two agents in an interaction are not two separate elements; rather, they influence each other, modulate one another's behavior, and engage in active co-regulation.

The actions of others are stimuli with effects on our own system. The relevance of the observed action either for the individual or as a shared goal modulates the action's effects. If the action is not relevant, it nonetheless activates synchronization processes that enable continuous monitoring and an adaptive response to the changing social environment (made up of objects and subjects). In addition, the observed but non-relevant action allows an automatic synchronization with the other, leading to a partial correspondence that facilitates understanding (co-cogitation) and adaptive co-action (co-regulation). If the action is relevant, it activates a goal-mediated synchronization consisting of automatic and controlled processes that promote selective responses to significant features of its dynamic context and which foster complementary actions. Automatic and controlled processes both are modulated by inhibitory and excitatory influences deriving from the context, and they jointly shape the mental representations of the stimuli that will subsequently be translated into motor reactions. This reaction provides information to the other agent, thereby generating an iterative process, a continuous closing and re-opening of an ever-changing circle.

Each of the agents attributes to himself/herself and to the other the status of agent, and does so by enacting self-monitoring processes. But the experience of acting together in a joint activity, as also suggested by Seemann [31], is an experience of *we acting*, of *us* as a common cause, enjoying a *sense of acting together*. As also noted by Searle and Bratman (see Par. 6.2.1. and 6.2.2), this sensation of properly interacting, i.e., of being agents together and sharing agency, cannot result merely from the sum of our sense of agency, of the agentive stance we attribute to others and perhaps of the awareness of a shared goal. This *summative account* is implausible and has little explicative power. Moreover, it cannot account for the sense of joint control that characterizes joint actions: in doing them, we feel that we are able to exert immediate control not only over our action but also over those of others. Effectively, we can influence the behavior of others even if this control is neither absolute nor complete. Likewise, joint control cannot be explained simply by summing up the amount of control we are able to exercise over others and vice versa.

Taking into account the sharing of agency and joint control, when we interact with someone else we have to understand that person, recognizing him/her as active and as an agent able to act in several ways, including those that escape the joint control shaping the experience of joint agency. Joint agency is therefore a complex and dynamic phenomenon, since it is more than the sum of the contributing individual agencies. Seemann [31] proposed the existence of a normative element to joint agency: "I expect that my co-agent will execute his/her part in our joint activity with continuity and competence or skill." This element does not derive from the sophisticated processing of thought and a prerequisite for joint activities. Instead, it consists of a primitive sharing of feelings and of other embodied intentional attitudes [31], similar to Searle's collective awareness but more evolved and more complex. This was defined by the author as *basic trust*, which creates and justifies the joint-agency experience. It is a non-cognitive phenomenon that can be described as an awareness of the potential for acting together efficiently.

It might be plausible, however, to pinpoint a second level of joint agency, thus drawing a parallel with the two-step account for agency proposed by Synofzik, Vosgerau, and Newen [32]. In line with the cognitive higher functions and propositional thinking distinctive of human beings, the embodied feelings referred to by Seemann could be involved in further processing, as they deal with the creation of an explicit judgment of joint agency. Imagine that you are playing tug-of-war with a friend against two other people, and after a while the other two fall and you and your friend win. At some point, the question will rise: are you the cause of the others' falling? Or does the cause have to be searched for in the context of the game, i.e., they simply slipped? Or is it possible to attribute the cause to them, i.e., they decided to stop playing and then, due to their diminished efforts, fell down? Thus, you could ask: Who caused that event? Assuming that the other team did not abandon the fight, was I the cause? Or was it my partner? Or was it our team?

Scenarios such as these focus our attention on an interesting but as yet unexplored issues of joint agency. By bringing real, everyday actions into the realm of research and the theorization of agency, we might enhance our understanding of joint action and thus of the role of agent.

6.5.1
The Intersubjective Origins of Joint Agency: A Developmental Perspective

We now return to the concepts of basic trust and joint agency. According to Chisholm [33], the embodied action experience may be seen as a necessary condition to develop and engage in common-cause thinking, i.e., the ability to think of ourselves as the common cause of many events and as having causal power in the environment. Seemann [31] postulated that the common-cause experience leads to an understanding of ourselves as agents, because it is *me* and in particular *my actions* that shape the context.

The experiences of acting and, more interestingly, joint-acting form the basis of agency and of joint agency, respectively, and begin in infancy. Accordingly, it could be useful to draw a parallel between the development of intersubjective perspective and the agentive stance. As Seemann observed [31], infants' capability to act jointly may be a consequence of their developmental, primary ability to commit themselves to episodes of mutual attention with a caregiver. After becoming able to manage this dyadic interaction, the infant develops the ability to jointly attend to third objects with others, thus becoming attuned to the inter-agent properties of his/her body and mind and the ability to share attention. This joint-attention ability then allows the development of and engagement in joint actions. Similar to our capacity for intersubjectivity, our skill to share feelings in episodes of mutual attention is the starting point for the development of the ability to attune to the embodied psychologies of others and to share mental states or attribute them to others.

In this perspective, the awareness of self as both an individual and an agent emerges from the experience that our own feelings can lead quite regularly to changes in social context and vice versa. Thus, the awareness of self and of the other as distinct identities in an interaction derives from the intersubjective stance, an embodied perspective in which sharing is not necessarily simple and does not mean mutual access to the mental lives of others (for a discussion of these concepts, see Par. 6.6). We can therefore hypothesize that direct sharing might extend to the dimension of action-related intentions, giving rise to a complete sense of joint agency and control during joint actions.

Sebanz [34] suggested an integrative view, hypothesizing that the sense of agency developed because basic kinds of social interaction rely on the capability to distinguish the action effects caused by oneself from those caused by others. Early in their development, human beings begin to enact basic forms of social interaction that do not depend on symbolic communication or high-level metacognitive functions. Through experiencing joint action, an individual develops a sense of self in action, supported by the adaptivity principle. In the course of evolution, the ability to distinguish between one's own capabilities and those of others conferred important evolutionary advantages. In this scenario, the first interactions would have been incidental, but as the individual registered that in the joint condition some effects can be achieved more easily than when attempted alone, the first distinction, acting alone *vs* not acting alone, was made. Then, based on this primitive sense of self, complementary action became available: by observing others, humans might have realized that others have similar action capabilities. But how can the individual make others do what he/she wants? This question would have brought about the understanding that action requires that others see what the individual sees, and thus the skills of drawing attention and developing joint attention. Finally, individuals would have noticed that the others are similar to themselves not only with respect to action and perception, but also in terms of desires and goals, leading to implementation of intentional imitation. Thus, humans memorized contingencies, developing a more stable sense of self and of the self as an acting entity with causal power among other, similarly acting entities.

6.6
Inter-acting Selves, Social Agency, and Neural Correlates

Although the sense of agency represents a useful concept within the social psychology and social neuroscience framework, its individual nature has limited investigations to those focused mainly on a personal, monadic level. However, it must be kept in mind that, according to Gallagher and Zahavi [5], while we are conscious of being the author of our actions, this awareness often occurs at the point when my actions are reflected by the presence of others: I become aware of myself through the eyes of other people. The more basic and direct contribution of agency to psychosocial theorization is inherent to the self-other differentiation. The capability to distinguish myself from the other is the first step in the evolution of social and interactive skills. Only if I recognize the other as such can I actually act upon him/her or with him/her and even ascribe to him/her specific attributes, such as mental states (e.g., intentions, desires, beliefs).

Many studies have investigated the mechanism underlying the attribution of agency, and neuroscience and brain imaging have identified its neural correlates. Thus, in the following paragraph, we discuss the neural structures pinpointed by empirical studies on agency and the possible anatomical-functional correlates of the self-other distinction in action and interaction.

6.6.1
The Original Distinction of Our-selves and Other-selves

Coherence between an internal sensory signal (i.e., somatosensory and proprioceptive), distal perceptual signals (i.e., visual and auditory), prior and situated intentions, and motor commands fosters and gives rise to a sense of agency. All these elements must be included in structuring a computational model of motor control, such as the *comparator model* proposed by Wolpert [35] (see also Chapter 1). In that model, the distinction between self-generated and externally generated actions is computed by the second comparator, coupling the predicted sensory state to the actual one. It is thus assumed that the sense of agency is associated with predicted-actual discrepancies evaluation.

Empirical brain imaging evidence acquired with functional magnetic resonance imaging (fMRI) and positron emission tomography (PET) suggested that the posterior parietal cortex, especially the inferior region, is involved in the monitoring of concordance between self-produced action and its visual consequences, detecting visuomotor incongruencies [36-39]. Also, the cerebellum may play a role in the signaling of discrepancies in predicted-actual effects. Based on real and virtual patient data from lesion studies and studies employing transcranial magnetic stimulation (TMS), respectively, it has been argued that the right inferior parietal cortex, and especially the temporo-parietal junction, plays a fundamental role in the attribution of movement agency and action awareness [40-43].

In a classical study, Spence and colleagues [44] used PET to identify the brain regions associated with experience of delusion of control; that is, an experience in which the patient feels that the actions (simple or complex), thoughts, or emotions he/she enacts, thinks, or undergoes are imposed by an exogenous force or controlled by some external agent rather than by self will [45] (see awareness [40-43] and Chapter 10). Similarly, interesting and useful conclusions were drawn from a study of schizophrenic patients. In the neurocognitive theory of schizophrenic delusional manifestations [46], passivity experiences are associated with abnormalities in predicting the sensory effects of an action, causing dysfunction of the prediction-actual state comparator module.

Spence scanned the brains of schizophrenic patients with and without delusional symptoms during the execution of a response selection task. The presence of delusions of control was associated with overactivity in the right inferior parietal cortex, interpreted as reflecting a heightened response to the sensory consequences of self-generated movements. The authors suggested that this heightened response may be related to a failure in modulating sensory areas based on effects prediction. Thus, both the inferior posterior parietal cortex and the cerebellum seem to be plausible neural correlates of the comparator mechanism involved in agency attribution [47].

Further interesting observations regarding the impairment associated with schizophrenia, were made by Frith, Blakemore, and Wolpert [48]. These authors suggested that the delusional patient, who lacks fundamental problems with action control, may not present abnormalities in the overactive parietal cortex but rather in the system that normally modulates activity in that site. The *anterior cingulate cortex*, involved in attention to future actions, may contribute to modulating the processes that underlie awareness of a system's predicted state. Connective streams, bringing information from one cerebral area to one or more other such areas, and thus contributing to cerebral activity coordination, are primarily involved in this type of modulating function. Dysfunctions related to cortico-cortical connections have been shown in patients with schizophrenia [49]. Studies using diffusion tensor imaging, a technique offering the opportunity to specifically explore interconnecting streams, could, in conjunction with fMRI findings, provide evidence for a role of neuronal connections in attribution processes and for white matter functionality.

6.6.2
Self-other Differentiation, Agency and Sociality: Hypotheses and Neuropsychological Evidence

Social situations imply at least the presence of two individuals, i.e., two potential agents, such that social attributes derive from the presence of the other, who is both a spectator and likewise an agent. Apart from attributing action to ourselves, we also attribute actions to others and this is a result of self-other differentiation. Evidence of a cognitive and neural overlap between self and other action representations led to the *who system* hypothesis. Georgieff and Jeannerod [50] suggested that distinguishing between one's own actions and someone else's actions is associated with the activation of non-overlapping brain areas. The prefrontal and parietal cortex, supplementary motor area, and cerebellum seem to be deeply involved in action attribution processes [14, 51], with the activity of the anterior insula and the right inferior region of the parietal lobe then modulated by the processing of performance-feedback discrepancies [25, 37, 38]. In this model, self-attribution of action is associated with greater activation of the anterior insula, while the attribution of agency to others activates the inferior right parietal cortex, and self-monitoring mechanisms related to error detection and feedback evaluation are associated with activation of rostral and caudal regions of the cingulate cortex.

As already stated, the possibility and capability to distinguish between self and other in agency attribution are essential requirements for acting in the world and with the external world. Notwithstanding the mainly active or passive role of an agent, during an interaction we need to be able to perceive our body, our acting space, and our causal power. Moreover, we have to perceive the other as an active system able to modify the context of interaction and to take part in an actual joint interaction (as stated in Par. 6.5).

In order to achieve a common goal, such as creating a house of cards, we can imitate the actions of others or coordinate our actions with theirs. Cooperation implies the integration of each subject's intentional plan, besides the sensorimotor one. Empirical evidence has given rise to a hypothesis regarding the existence of a circuit of cortical areas (other than the putative MNS) that is involved and activated during complementary joint action. A superadditivity criterion has been suggested, similar to that of Georgieff and Jeannerod and based on overlapping cerebral activation patterns [27].

In light of empirical data, Kokal, Gazzola, and Keysers [27] proposed a *dual process model* for joint-action management, therefore including agency attribution. The models states that observed actions are firstly transformed into representations of similar actions in the observer through a combination of the forward and inverse models we develop while observing the consequences of our own actions and those of others and while preparing to act. These series of operations are associated with a set of brain areas, including the putative MNS, which ensures that the two essential components that need integration during joint actions, our own actions and those of others, are in the same neural code. This code can be predominantly motor or sensory or a hybrid and is located in different regions of the brain (premotor cortex, superior temporal sulcus, or parietal cortex). Those commonly coded representations and prepared behavioral alternatives are finally processed by a second set of areas adjacent to MNS regions and showing additional activity during integration in joint actions. The results of processing are self-other action differentiation (and thus recognition of self and other as agents), but also flexible integration of our own intentions and actions with those of others, as well as the selection of the most adequate action, amongst the alternatives primed in the premotor cortex, to achieve our chosen joint goal.

6.7
Conclusions

We have addressed agency question in the framework of its individual and social aspects, outlining a path from initial remarks on mental elements in collective action to an explicit discussion about interpersonal domain and the possible origins of agency and joint-agency. Agency, as we have seen, is a complex phenomenon that includes intention, goals, and desires but also somatosensory signals and body schema. The computational comparator model contains all the low-level and high-level elements that describe agency and it foresees their integration in order to promote action awareness and action control. The integration of these different levels is necessary also for engendering a complete agency experience.

As we have seen, agency-related processes strictly interweave with sociality and constitute a bridge between two of its basic functions (studied in two related research fields in social neuroscience): *social perception* and *social cognition*. The former deals with the initial stages in the processing of information that culminate in the accurate analysis of the dispositions and intentions of other individuals [52]; it is based, firstly, on a pre-conceptual attribution of an agentive stance. The latter involves the ability to construct representations of the relations between oneself and others and to use those representations flexibly to guide social behavior [53]; it is related to self-other differentiation and to the management of this differentiation in order to achieve efficient joint action.

The link between action and self-other agency is an important element in the definition and management of social interaction. Only if I feel that I am an agent can I activate self-monitoring, develop an intentional plan, or define a joint strategy; only if I recognize the other as another agent can I can actually interact with him/her.

The concept of agency is certainly deeply rooted in the individual dimension. However, besides the "classical" social aspects of agency, for example attributional function, it may be possible to broaden the analysis so that it includes complex ecological contexts, by exploring the features of the sense of agency within distributed collective actions and joint actions, both imitative and co-operative ones.

References

1. van den Bos E, Jeannerod M (2002) Sense of body and sense of action both contribute to self-recognition. Cognition 85:177-187
2. Gallagher S (2000) Philosophical conceptions of the self: implications for cognitive science. Trends Cogn Sci 4:14-21
3. Bandura A (1986) Social foundations of thought and action: a social cognitive theory. Prentice-Hall, Englewood Cliffs
4. Newen A, Vogeley K (2003) Self-representation: searching for a neural signature of self-consciousness. Conscious Cogn 12:529-543
5. Gallagher S, Zahavi D (2009) Phenomenological approaches to self-consciousness. In: Zalta EN (ed) The Stanford Encyclopedia of Philosophy, Spring 2009 ed., http://plato.stanford.edu/archives/spr2009/entries/self-consciousness-phenomenological/
6. Searle JR (1990) Collective intentions and actions. In: Cohen PR, Morgan J, Pollak ME (eds) Intentions in Communication. MIT Press, Cambridge, pp 401-415
7. Bratman ME (1992) Shared cooperative activity. Philos Rev 101:327-341
8. Crivelli D, Balconi M (2009) Trends in social neuroscience: from biological motion to joint actions. Neuropsychol Trends 6:71-93
9. Knoblich G, Sebanz N (2006) The social nature of perception and action. Curr Dir Psychol Sci 15:99-104
10. De Jaegher H (2009) Social understanding through direct perception? Yes, by interacting. Conscious Cogn 18:535-542
11. Gallagher S (2008) Direct perception in the intersubjective context. Conscious Cogn 17:535-543
12. Meltzoff AN, Moore MK (1997) Explaining facial imitation: a theoretical model. Early Dev Parent 6:179-192
13. Gallese V, Goldman A (1998) Mirror neurons and the simulation theory of mind-reading. Trends Cogn Sci 2:493-501
14. Grèzes J, Decety J (2001) Functional anatomy of execution, mental simulation, observation, and verb generation of actions: a meta-analysis. Hum Brain Mapp 12:1-19
15. Annet J (1996) On knowing how to do things: a theory of motor imagery. Cogn Brain Res 3:65-69
16. Meltzoff AN (1999) Origins of theory of mind, cognition and communication. J Commun Disord 32:251-269
17. Pacherie E, Dokic J (2006) From mirror neurons to joint actions. Cogn Syst Res 7:101-112
18. Sebanz N, Bekkering H, Knoblich G (2006) Joint action: bodies and minds moving together. Trends Cogn Sci 10:70-76
19. Blakemore S-J, Decety J (2001) From the perception of action to the understanding of intention. Nat Rev Neurosci 2:561-567
20. Blakemore S-J, Frith CD (2005) The role of motor contagion in the prediction of action. Neuropsychologia 43:260-267

21. Fadiga L, Craighero L (2004) Electrophysiology of action representation. J Clin Neurophysiol 21:157-169
22. Léonard G, Tremblay F (2007) Corticomotor facilitation associated with observation, imagery and imitation of hand actions: a comparative study in young and old adults. Exp Brain Res 177:167-175
23. Sebanz N, Knoblich G, Prinz W (2005) How two share a task: co-representing stimulus-response mappings. J Exp Psychol Hum Percept Perform 31:1234-1246
24. Sato A, Yasuda A (2005) Illusion of sense of self-agency: discrepancy between the predicted and actual sensory consequences of actions modulates the sense of self-agency, but not the sense of self-ownership. Cognition 94:241-255
25. Balconi M, Crivelli D (2010) FRN and P300 ERP effect modulation in response to feedback sensitivity: the contribution of punishment-reward system (BIS/BAS) and behaviour identification of action. Neurosci Res 66:162-172
26. Keysers C, Fadiga L (2008) The mirror neuron system: new frontiers. Soc Neurosci 3:193-198
27. Kokal I, Gazzola V, Keysers C (2009) Acting together in and beyond the mirror neuron system. Neuroimage 47:2046-2056
28. Newman-Norlund RD, Bosga J, Meulenbroek RGJ et al (2008) Anatomical substrates of cooperative joint-action in a continuous motor task: virtual lifting and balancing. Neuroimage 41:169-177
29. Newman-Norlund RD, van Schie HT, van Zuijlen AMJ et al (2007) The mirror neuron system is more active during complementary compared with imitative action. Nat Neurosci 10:817-818
30. Semin GR, Cacioppo JT (2009) From embodied representation to co-regulation. In: Pineda JA (ed) Mirror neuron systems. Humana Press, New York, NY, pp 107-120
31. Seemann A (2009) Joint agency: intersubjectivity, sense of control, and the feeling of trust. Inquiry 52:500-515
32. Synofzik M, Vosgerau G, Newen A (2008) Beyond the comparator model: a multifactorial two-step account of agency. Conscious Cogn 17:219-239
33. Chisholm R (1995) Agents, causes and events: the problem of free will. In: O'Connor T (ed) Agents, causes, and events: essays on indeterminism and free will Oxford University Press, New York, pp 95-100
34. Sebanz N (2007) The emergence of self: sensing agency through joint action. J Conscious Stud 14:234-251
35. Wolpert DM (1997) Computational approaches to motor control. Trends Cogn Sci 1:209-216
36. Chaminade T, Decety J (2002) Leader or follower? Involvement of the inferior parietal lobule in agency. Neuroreport 13:1975-1978
37. Farrer C, Franck N, Georgieff N et al (2003) Modulating the experience of agency: a positron emission tomography study. Neuroimage 18:324-333
38. Farrer C, Frith CD (2002) Experiencing oneself vs another person as being the cause of an action: the neural correlates of the experience of agency. Neuroimage 15:596-603
39. Fink GR, Marshall JC, Halligan PW et al (1999) The neural consequences of conflict between intention and the senses. Brain 122:497-512
40. McDonald PA, Paus T (2003) The role of parietal cortex in awareness of self-generated movements: a transcranial magnetic stimulation study. Cereb Cortex 13:962-967
41. Preston C, Newport R (2008) Misattribution of movement agency following right parietal TMS. Soc Cogn Affect Neurosci 3:26-32
42. Sirigu A, Daprati E, Ciancia S et al (2003) Altered awareness of voluntary action after damage to the parietal cortex. Nat Neurosci 7:80-84
43. Sirigu A, Daprati E, Pradat-Diehl P et al (1999) Perception of self-generated movement following left parietal lesions. Brain 122:1867-1874

44. Spence SA, Brooks DJ, Hirsch SR et al (1997) A PET study of voluntary movement in schizophrenic patients experiencing passivity phenomena (delusions of alien control). Brain 120:1997-2011
45. Blakemore S-J, Wolpert DM, Frith CD (2002) Abnormalities in the awareness of action. Trends Cogn Sci 6:237-242
46. Frith CD (1992) The cognitive neuropsychology of schizophrenia. Lawrence Erlbaum, Hove
47. David N, Newen A, Vogeley K (2008) The "sense of agency" and its underlying cognitive and neural mechanisms. Conscious Cogn 17:523-534
48. Frith CD, Blakemore S-J, Wolpert DM (2000) Abnormalities in the awareness and control of action. Philos Trans R Soc Lond B Biol Sci 355:1771-1788
49. Fletcher P, McKenna PJ, Friston KJ et al (1999) Abnormal cingulate modulation of fronto-temporal connectivity in schizophrenia. Neuroimage 9:337-342
50. Georgieff N, Jeannerod M (1998) Beyond consciousness of external reality: a "Who" system for consciousness of action and self-consciousness. Conscious Cogn 7:465-477
51. Vogeley K, Fink GR (2003) Neural correlates of the first-person-perspective. Trends Cogn Sci 7:38-42
52. Allison T, Puce A, McCarthy G (2000) Social perception from visual cues: role of the STS region. Trends Cogn Sci 4:267-278
53. Adolphs R (2001) The neurobiology of social cognition. Curr Opin Neurobiol 11:231-239

Section III
Clinical Aspects Associated with Disruption of the Sense of Agency

Disruption of the Sense of Agency: From Perception to Self-knowledge

M. Balconi

7.1
Introduction

Different planes of disruption may result in a loss or breakdown of the sense of agency. They may pertain to the different levels of perceptual, attentional, and psychiatric disorders, and generally they have in common a loss of the *sense of control* and a disruption of *binding* between *intentions* and *actions*. In some cases, they imply a deficit in the sense of *ownership*, which may be subdivided into *ownership for body-actions* and *for thought-action*. In other cases, what is lost is the *sense of agency* itself, similarly divided into *agency for action* and *agency for thoughts*.

This chapter discusses disruption of the sense of agency, taking into account these different levels of analysis. Specifically, we consider perceptual deficits, such as blindsight and arm illusions, which are related to agency misidentification, as well as attentive deficits, which are related, by contrast, to disruption in attention allocation; for example, as in neglect syndrome. Finally, the psychiatric disturbances schizophrenia, autism spectrum disorder, ands obsessive-compulsive disturbs are considered as examples of psychological impairments in the perception of agency.

7.2
Disruption of Agency in the Perceptual Field and in Proprioception

Specific perceptual deficits are related to the representation of self within a given context and to the representation of one's own body. The first type of deficit is a lim-

M. Balconi (✉)
Department of Psychology, Catholic University of Milan, Milan, Italy

itation in self-recognition, such as in *anosognosia* and *somatophrenia*. These refer to a condition in which an individual who has sustained a brain injury resulting in paralysis is unaware of the weakness or loss of function in the paralyzed body part. The second is due to perceptual mis-identification of parts of the body. Both types of perceptual deficits are analyzed in the following.

7.2.1
Agency and Body: Predictivity Function of the Body for Self-representation

Body perception and body consciousness offer a rich descriptive basis for the characterization of embodiment, which, in turn, provides a starting point for theories of the self and of agency [1]. Throughout this volume, we have underlined that one of the main contributions to the sense of agency is derived from body representation. Bodily self-knowledge may be based on *visual information* or on *proprioceptive information*. Visual information includes a prior visual identification: since I can see my own body as well as the bodies of other people, I need to distinguish between mine and theirs. Proprioceptive information is identification-free and directly relies on my proprioceptive system: as I cannot receive any proprioceptive information about someone else's body, I am assured that the source of information is my own body. Consequently, proprioceptive self-ascription does not depend on the identification of the body as one's own, unlike the visual self-ascription of bodily properties.

Identification implies the possibility of mis-identification whereas representations that are identification-free are considered immune to this type of error. When I recognize the action of grasping a glass as my own, the sense of agency involves the notion of a *minimal self* [2] (see Chapter 3), which is instantaneous and which carries only one bit of information about myself, e.g., that I am grasping a glass. The knowledge that I am the person I see in the mirror, however, depends on recognitional criteria that allow me to re-identify myself through time. The "recognitional self" has a richer content and may even constitute personal identity.

Another matter of interest is the extent of the effective contribution of the body to the sense of agency. To address this issue, we need to introduce a clear distinction between body ownership and agency (see also Chapter 10). Whereas agency is the sense of intending and executing actions (including the feeling of controlling one's own body movements) and events in the external environment, body ownership refers to the sense that one's own body is the source of sensations. The sense of body ownership is present not only during voluntary actions, but also during passive experience; however, only voluntary actions produce a sense of agency. Thus, body awareness refers to proprioceptive awareness and, by extension, to the conscious experience of the location of a specific body-part in space [3].

7.2.2
Perceptual Illusions of Body

To explain the difference between body ownership and agency, we can experimentally manipulate body awareness during active movements and passive stimulation. The *rubber hand illusion* consists of watching a rubber hand being stroked synchronously with one's own unseen hand, which causes the subject to attribute the rubber hand to his or her own body, as if it is a proper (and the subject's) hand. Correlated visual and tactile information induces a changed awareness of one's own body, resulting, for instance, in the incorporation of the rubber hand into the body. The degree of incorporation can be measured quantitatively via the drift in proprioceptively perceived position.

One form of impairment in intentional action can be found in patients with *anarchic hand syndrome* [4]. These patients make hand movements that are not under their volitional control. However, the anarchic hand behaviors are accepted by the patient as being their own, even though not under the control of intention. This deficit may be due to a perceived mismatch between the inappropriate action and the goal of the patient. Patients find one of their hands performing complex, apparently goal-directed movements they are unable to suppress. Sometimes the anarchic hand interferes unhelpfully with intentional actions performed with the use of the other hand. Sometimes it performs movements apparently unrelated to any of the agent's intentions. In most of these cases, the patient goes on to claim that he feels as if the actions performed by the anarchic hand are not his, and that this hand is doing something that was not intended or wanted, and that cannot be controlled. The most immediate question that might come to mind about the anarchic hand's movement is whether it is even an action. The answer should be negative, since the patient is neither in control of nor at all responsible for the movement. Nevertheless, anarchic hand movements are also not pure reflexes; they are clearly devoted to a particular goal, and, relative to the goal, well-executed. What appears to be disrupted in this case is the causal explanation of actions, the mechanism of control, and the source of our knowledge of actions. Thus, anarchic hand patients have a sort of control and of awareness of the hand's action, but something is lacking with regard to the two processes.

A second type of disruption in the intentionality of acting is *alien hand behavior*, in which patients fail to report ownership of the wayward limb and may not know when an appropriate action has been made. This form of behavior might correspond to the disruption of attributional processing, and in pathological conditions it is an exaggerated form of what is observed in normal people.

Neuropsychological evidence suggests that in the case of the anarchic hand there is damage within one hemisphere to the neural region involved in the internal control of action (supplementary motor area) as well as lesions of the corpus callosum that disrupt communication from the unaffected hemisphere. Thus, hand actions made by the damaged hemisphere are driven by environmental factors rather than the patient's intention. Many studies found similar results in terms of the specific deficits related to this disturbance, in particular, that patients' responses were modulated by the task and by the familiarity of the objects as well as by the familiarity of their orientation [5].

Specific deficits related to delusions of control corroborate the suggested roles of some brain areas (parietal cortex and cerebellum) in anticipating future states of the body and distinguishing between self and other. In patients with delusions of control, active movements are processed in the brain as passive movements, and many of these patients report feelings of alien control while performing the willed-action task, demonstrating a correspondence between brain activation patterns and the phenomenology of passivity experiences.

7.2.3
Blindsight and Numbsense

A consistent amount of research has stressed the presence of a dissociation between conscious and non-conscious body representations. Anosognosic patients exhibit some form of unconscious knowledge about their bodily deficit. The most famous example of this type of dissociation is *blindsight*. The term itself illustrates the paradoxical nature of this phenomenon, described as cortical blindness. Patients with a lesion of the primary visual system do not perceive visual stimuli presented within the area of the visual field affected by the lesion. However, they remain able to move their eyes or hands toward a stimulus when accurately instructed [6]. The initial interpretation of blindsight was based on the sub-cortical *vs* cortical visual system, but successive theoretical contributions suggested that the condition involves the cortical visual pathway [2].

Numbsense, the somatosensory equivalent of blindsight, demonstrates that it is possible to isolate unconscious body representations that are highly specific to action. The case of a patient incapable of any somatosensory sensations from the forearm but able to point to the locus of stimulation applied to her unfelt forearm provides evidence of a clear dissociation between "where" and "what," a condition referred to as *blindtouch* [7]. One famous case involved a patient whose proprioceptive deficit was due to a left parietal thalamo-subcortical lesion [8]. He was unaware of any kind of somatosensory stimuli applied to his arm and leg and failed to demonstrate any significant performance in a verbal forced-choice paradigm. Nevertheless, he performed significantly well in tests requiring pointing to the tactile stimulus location on the numb arm.

An important issue in both blindsight and numbsense is whether the presence of residual sensory-motor abilities could be used as the basis for reverberating sensory information to the perceptual systems. It was shown that, when patients performed simultaneous motor and perceptual responses, they lost their residual motor abilities and their perceptual performance did not improve. This result suggests that the cognitive representation of the stimulus, once activated, systematically replaced the residual sensorimotor representation of the same stimulus. Thus, in both cases, the cognitive representations were hierarchically higher than the sensorimotor ones [9].

7.2.4
A Tentative Conclusion Regarding Perceptual Level Impairment

Behavioral and neuroimaging data support the hypothesis that impairment on the perceptual and attentive levels results from the integration of different signals. Specifically, the body schema, as a significant example of body representation, arises from the integration of different signals. In body representation processes it is possible to identify a progressive, hierarchical level of analysis, ranging from an elementary operation, such as somatosensory processing, to a more complex computation, such as the construction of the feeling of ownership of body parts. Nevertheless, the boundaries between these processes are unfocused. For example, the possibility of modulating apparently elementary deficits, such as tactile imperceptions, through physiological stimulations that allow transient conscious perception of the tactile stimuli suggests that even low-level deficits involve a higher component.

7.3
Attentive Deficits and the Sense of Agency

Attention may be impaired in the case of anomalous subjective representation of internal or external cues, which results in a common absence of response to background cues. In the following, we focus on two main forms of attentive deficit: visual neglect and somatosensory neglect.

7.3.1
Visual Neglect Syndrome

A specific syndrome was observed that revealed a systematic perceptual inability to process stimuli localized within the contralesional space (generally the left side, with right-hemisphere lesions). Unilateral neglect generally comprises failure to acknowledge, respond to, orient to, or report stimuli occurring on the left side of the patient's personal or extrapersonal space.

A type of lateral bias of spatial attention occurs in the context of a reduced attentional capacity [10], whereby stimuli on the ipsilesional side briefly attract the individual's attention, to the exclusion of simultaneous stimuli located in the contralesional space. Both unilateral neglect and extinction are dissociated from primary sensory losses such as hemianopsia, suggesting that the source of these disorders is at a higher information-processing level. Although evident primarily in the visual modality, neglect may also be manifested in the auditory and tactile modalities. Early theories suggested that neglect results from a deficit in the sensory or perceptual processing of neglected stimuli and is independent of primary sensory deficits, which may be manifested also in absence of an external stimulation. Also, impairments in attentional mechanisms were proposed to explain neglect phenomena, specifically, as a breakdown in a system that normally allocates attentional resources to locations in the neglected hemisphere. According to recent models, neglect reflects a failure of the *representational system* that maps the external space into a neural system. This hypothesis was supported by results indicating that neglect patients seem to neglect the left parts of space even in their imaginations [11].

There is no doubt that overt behavior is affected by neglect, as is the self-perception of these patients' efficacy in perception or in action execution related to the contralesional space. More generally, neglect is considered as a deficit of conscious access to information coming from the contralesional side of space. Here we examine the main deficits of neglect, those related to space and body representation. External neglect is manifested in stimuli delivered in any sensory modality, although it is mainly studied in the visual domain. Patients appear to omit items presented to their left and may exhibit sustained eye and head deviations to the right, estimating the straight-ahead direction to be shifted to the right. In addition, their body image can be affected. For example, they may show anosognosia, that is, a lack of awareness for, e.g., a left-sided deficit such as hemiplegia, or even somatoparaphrenia (delusions about their own body). Several studies have shown that these patients are more impaired in perceptual than in visuomotor tasks. They may be strongly biased when requested to indicate the middle of a stick with their fingers, but they are relatively less impaired when the task is simply to grasp the object [12]. This finding suggests that in unilateral neglect lower-level visuomotor functions are relatively spared.

Moreover, patients with unilateral neglect may exhibit a deficit of the egocentric reference frames used for action and for self-body representation. These patients generally have difficulties to represent their own body, as their left half is strongly neglected. Higher-order deficits of this kind contrast with the preserved visuomotor abilities and are compatible with the view that neglect is primarily a deficit of conscious access and use of information [13].

7.3.2
Somatosensory Neglect

Some aspects of the mutual interactions that contribute to the feeling of having a body acting in space are becoming better understood due to progress in identifying the neuropsychological correlates of the different levels of body representation in normal subjects and in patients with disorders in body representation. For example, somatosensory neglect, resulting from lesions of the somatosensory parietal cortex, may induce impairments in reporting tactile stimuli delivered to the contralesional side of the body. Left somatosensory deficit may reflect not only a primary sensory impairment but also a more complex and higher-order deficit of spatial representation of the body [14]. The observation of behavioral asymmetry (more left than right deficit) has suggested that this form of hemianesthesia includes an attentional-spatial component strictly related to the neglect syndrome.

As previously stated, unilateral deficits are typically associated with right-brain damage. Thus, it is very common that patients with left-side neglect present a wide variety of body schema impairments (for the concept of body schema, see [15]). The symptoms include personal neglect, which is the inability to orient toward, explore, and perceive the contralesional half of the body (anosognosia). Sometimes neglect patients do not use the contralesional limb even in the absence of primary motor deficits such as a left hemi-paresis. A few cases have been observed in which patients with personal neglect do not also exhibit extrapersonal neglect [16]. The evidence for the behavioral dissociations associated with neglect suggests that the representation of different parts of space (extrapersonal, peripersonal, *vs* personal, bodily space) may be subserved by functionally dissociated and independent systems.

While the anatomy of peri- and extrapersonal neglect has been extensively studied [17], the anatomical substrate of personal neglect is poorly understood. Nevertheless, it seems that personal neglect is associated with posterior brain lesions involving infero-posterior parietal areas or subcortical regions such as the basal ganglia, thalamus, and white-matter fiber tracts.

Patients affected by some form of body representation disorder may not acknowledge their deficits despite unambiguous evidence. Anosognosia is, in some cases, not specific but selective (i.e., only for a limb or some kind of movements), which implies that it cannot be explained as a generalized disturbance of awareness (for example, related to damage to prefrontal areas). The selectivity of anosognosia suggests that awareness has a composite structure, revealing even at the level of thought structure the modular organization of the cognitive system. Recently, it was shown that its occurrence is linked to damage to the frontal and parietal lobes. The authors suggested that anosognosia can be viewed as a disorder of motor awareness implemented in a fronto-parietal circuit related to space and motor representation, in which the parietal component may be responsible for the spatial computation necessary to act in space [18].

7.4
The Fallibility of Self-attribution of Agency in Neuropsychiatry

The claim of ownership, the self-ascription that I am the subject who is undergoing an experience, can be consistent with the lack of a sense of agency. Phenomena such as delusions of control and thought insertion appear to involve problems with the sense of agency rather than with the sense of ownership. More generally, pathological cases of sense of control for thoughts can be classified in three main categories: (1) utilization behavior, in which subjects apparently lose the ability to overcome the power of a stimulus to invoke a habitual action; (2) perseveration, in which subjects seem to be unable to stop performing a particular sequence of actions; and (3) thought insertion, in which subjects are convinced that thoughts are being inserted into their minds by external forces. Later on, we consider the last case in order to explain the failure of control by an agent.

7.4.1
Frontotemporal Dementia and the Delusion of Control in Frontal Deficits

There is agreement on the fact that injury to the inferior medial frontal area results in a loss of inhibition and thus in the increased expression of behaviors considered nonsocial, while injury to the more superior medial frontal lobe results in a more apathetic state. Injury to this area of the brain is also typical of fronto-temporal dementia (FTD), a progressive neurodegenerative deficit beginning in mid- or later adulthood in which personality change or language disorder develops as an initial symptom of a progressive global dementia [19]. In FTD, personality and personal identity are altered but these individuals are usually unaware of the changes and reject the claim that such changes have taken place. This type of dementia differs significantly from Alzheimer's disease, since it is a degenerative condition primarily affecting the frontal and the anterior temporal lobes, areas that control judgment, personality, behavior regulation, speech and social interactions, and some aspects of memory.

Some patients with frontal lesions automatically execute the action performed by someone else that they are observing, losing track of the distinction between their own intentions and the intentions of others. In general, imitation constitutes a bridge that carries interpersonal information and plays a major role in interactions between people, starting from birth [20]. Imitation implies innate mapping from self to others and can be understood only if we postulate the existence of shared representations of actions between the imitator and its target. We emphasize the commonality between self and others. However, since a mirror matching mechanism has been found in humans for self- and other-generated actions [21], how can the subject discriminate between internal and external sources of the activated representation? There must be a mechanism that enables us to self-attribute our own actions, that is, a system that allows the subject to know without ambiguity who the agent is even in complex situations such as mutual imitation. But this mechanism is not infallible; in fact, its sensitivity is limited even in normal subjects and it is possible to self-attribute movements that are different from those actually performed by oneself. Moreover, the very existence of such a mechanism implies the possibility of its breakdown, as is the case in positive symptoms of schizophrenia.

7.4.2
Agency and Schizophrenia

The sense of ownership for motor action can be explained in terms of the ecological self-awareness built into movements and perceptions. By contrast, experimental research on normal subjects suggests that the sense of agency for action is based on that which precedes action and translates intention into action. Taking into account the comparator model (see Chapter 1), if a forward mechanism fails, sensory feedback may still produce a sense of ownership but the sense of agency will be compromised, even if the actual movement matches the intended movements. Schizophrenic patients most likely have problems with this forward, pre-action monitoring of movements but not with motor control, based on a comparison of intended movement and sensory feedback.

The neuroscience of action and the neuropsychology of schizophrenia confirm the existence of specific cognitive processes underlying the sense of agency, the "who" system [22], which is responsible for the critical distinction between oneself and the other and is disrupted in delusions of control [23]. Thus, the "who" system allows self-attribution, with the latter taking place in a social frame of reference of shared representations between the self and others. One class of symptoms displayed by schizophrenic patients seems to be closely related to dysfunction of the "who" system. These symptoms include thought insertion, auditory-verbal hallucinations, delusions of reference, and delusions of alien control. By impairing the distinction between the self and the external world, these false beliefs lead to a feeling of depersonalization.

During verbal hallucination, schizophrenics talk to themselves but are unaware of doing so. Similarly, those with delusions of control may believe that they control the actions performed by someone else or that their own actions are influenced by the will of other people. Neuroimaging studies in schizophrenic patients provide interesting information with which to interpret these symptoms. During verbal hallucinations, there is abnormal activation of primary auditory cortical areas: the patient hears his inner speech as if it were the voice of someone else. A similar phenomenon is observed during the generation of spontaneous arm movements. Brain activation was found to be increased in a cortical network including the left premotor cortex and the right inferior parietal lobule. This result, together with those obtained for verbal hallucinations, can be interpreted as indicative of a deficit in cortico-cortical inhibition, which normally suppresses activity in critical areas during self-produced action. Lack of inhibition in these regions would therefore lead to incorrect agency judgments, with a tendency to misattribute actions to an external agent.

Positive schizophrenic symptoms, especially passivity phenomena and including auditory hallucinations, may be caused by an abnormal sense of agency— often exhibited by people with schizophrenic personality traits. It is possible that this abnormal sense of self-agency is attributable to the abnormal prediction of one's own movements in motor control. A recent empirical contribution comes from an experiment using the "disappeared cursor" paradigm, in which non-clinical, healthy participants and those with schizophrenic personality traits were required to click on a target using an invisible mouse cursor [24]. Prediction error was defined as the distance between the target and the click point. The results showed that people with schizophrenic personality traits correlated with deficits in predicting movements of their left hand. In particular, auditory hallucination proneness had the strongest relationship with movement prediction error. In this context, we should also discuss error tendency (overestimations or underestimations of one's own movements), as it is in accordance with the idea that passivity phenomena or proneness is caused by the abnormal prediction of one's own actions or movements.

The sense of agency in "I thoughts" has been examined mainly as part of investigations into thought insertion in schizophrenia. Based on an analysis of this condition, in the healthy individual, a coherent and plausible story about his or her own mental life seems to take into account various factors, such as background beliefs, concordance with the self-image and the line of thoughts, and cognitive effort and is generated by some sort of rationalization module. Disruption of this module may produce a strong sense of impersonal thoughts or of the external insertion of thoughts [25].

7.4.3
Concluding Remarks on Schizophrenia

How does it happen that one is able recognize oneself as the source of one's own actions? We have observed that this process of self-recognition is far from trivial: although it operates covertly and effortlessly, it depends upon a set of mechanisms involving the processing of specific neural signals, of sensory as well as central origin. In a recent study, experimental situations in which these signals were dissociated from each other such that self-recognition became ambiguous were applied in healthy subjects and in schizophrenic patients. The results revealed that there are two levels of self-recognition, an *automatic level* for action identification, and a *conscious level* for the sense of agency, both relying on the congruence of action-related signals. The automatic level provides an immediate signal for controlling and adapting actions to their goal, whereas the conscious level provides information about the intentions, plans, and desires of the author of these actions. In schizophrenic patients, these two levels can be dissociated from each other such that whereas automatic self-identification is functional in these patients, their sense of agency is deeply impaired. The predominant symptoms, which represent one of the major features of the disease, testify to the loss of the ability of schizophrenic patients to attribute their own thoughts, internal speech, and/or covert or overt actions to themselves.

Application of the comparator model to schizophrenia confirmed experiments showing that schizophrenics tend to rely more on external action signals (for example, visual feedback) than on internal cues (proprioception and efference copy) when perceiving their own actions and inferring agency. This may explain why schizophrenics also self-attribute external sensory events to their own actions even if these actions are largely manipulated. If the impairment lies in an inadequate integration of different authorship indicators, the break-down in self-monitoring might be readily explainable without postulating a deficit in the comparator.

Indeed, many dysfunctions in perceptual processing that do not need to postulate a deficit in the comparator are conceivable. For example, one study found that when patients with delusions of persecution were observing animation sequences, they attributed intentional behavior to the moving shapes whereas controls attributed no intentionality [26]. The failure of agency attribution might also result from a nonspecific failure in monitoring internal cognitive sources in tasks that do not use any action control or action perception at all. Moreover, a failure to disambiguate complex signals on the basis of accurate inferences from context situations can often be relevant.

7.4.4
Autism: Mentalizing *vs* Agency Disruption

Explorations of autism spectrum disorder (ASD) have considered the many key components of social cognition, such as detection of animacy [26], perspective taking [27], the ability to engage in meaningful imitation [28], joint attention [1], and the sense of agency [3, 29]. These constructs were considered as precursors of the ability to *mentalize* (attribute mental states to others) or of *theory of mind* [1, 30]. While mentalizing requires the differentiation of one's own mental state from those of others, agency requires the differentiation of one's own actions from those of others.

Predictions on difficulty with the sense of agency in ASD can be made based on reported deficits in imitation (DSM-IV), motor performance, executive functioning, multimodal integration, and temporal binding. From a neuropsychological point of view, the cerebellum and frontal lobes, both of which have been implicated in action monitoring, were observed to be impaired in some autistic subjects [31]. Nevertheless, a large body of empirical work regarding the social deficits in autism has focused on mentalizing and imitation. These properties are associated with simulation theory but little work has been done on the self-other distinction or on the sense of agency.

The contribution of the sense of agency, and its possible disruption, to autistic behavior was explored using a task based on the results of previous agency-manipulation studies [32, 33]. Comparing the deficits of autistic subjects in mentalizing and in agency representation, the authors were able to analyze in greater detail the effect of these different components on ASD. Their results helped to elucidate the relation between the two processes, since they found that socio-cognitive difficulties in autism occur on a level higher than that needed for action monitoring and awareness of action. The data suggested, in fact, that the problems that autistic people have with imitation are not grounded in deficient action monitoring or awareness of their own or others' behavior. Moreover, as in the case of simulation deficit, the sense of agency is preserved, since autistic subjects showed deficits in a mentalizing task, engaging simulation, but did not show agency deficits, suggesting that simulation does not underlie the sense of agency.

An alternative explanation of this result may be that agency and mentalizing are two independent, unrelated processes that are based on distinct neurocognitive mechanisms [34]. However, contrasting evidence to this interpretation comes from schizophrenia, in which both processes are impaired. People with ASD rarely show positive symptoms or a disintegration of personality, as manifested in mis-attributions of agency or delusions of control. Thus, alternatively, the dissociation between the sense of agency and mentalizing in ASD suggests that the former involves a preconceptual aspect while the ascription of mental states to others is metarepresentational. It has been emphasized that agency may be represented from two different perspectives: a lower-level, preconceptual, and sensorimotor level (feeling of agency) and a higher (metarepresentational), conceptual level (judgment of agency) (see also Chapter 3) [35]. The problems of ASD subjects are thought to involve the latter.

7.4.5
Dissociated States: Obsessive-compulsive Disorder

Obsessive-compulsive disorder is considered a significant pathological profile that includes many symptoms related to action planning and action execution. Obsessions can be interpreted as intrusive and uncontrolled thoughts, whereas a compulsion is the urge to perform stereotyped mental or physical actions. Compulsions may be associated also with discomfort regarding particular sensations of incongruity or failure, sensations of incorrectness, and feelings of imperfection [36]. Repeated actions are thought to represent the inability to formulate a sense of task completion, regardless of goal attainment. Recent research has found obsessive-compulsive behavior to be related to deficits affecting action processing, specifically, *action planning*, in terms of defective use of internal representations to guide action, and *action monitoring*, as a conflict between external outcomes of actions and internal representations, in which the agent detects an "error signal" and tries to correct actions. More generally, a defective representation of goal-directed habitual action is present. Usually, the actions are mentally represented, with a correct link between a goal and the instruments used to realize it, and they are monitored according to their internal, related goal. By contrast, performing actions with unavailable or meaningless goal representations may produce disorganization in action flow [37].

A general representation of action and of the relationship between means and goals with respect to action representations is proposed by the *action identification theory* (AIT) [38]. According to the AIT, any behavior may be identified within a cognitive hierarchy of meanings, in which the lower-level represents instrumental features, and the higher-level relates to the desired goal of the action. The specific level at which action is represented may reflect the accessibility of a particular representation: higher-level action identification is usually used to perform routine and well-known actions, and lower-level identification to perform recently learned actions. In the case of complex actions or a disruption of action, low-level identification tends to be adopted. The level of agency refers to the preferential level at which actions are generally identified, which reflects the internal representation (goal *vs* movement) that is generally activated during an action. Distinct modes of action are promoted by the different levels. In the absence of goal representations, gestural representations guide actions, with chronic low-level identification thought to promote signals of inconsistency and error during routine actions.

Moreover, obsessive-compulsive disorders can be characterized by a specific *style of action representation*, i.e., the level of personal agency. There is, also, a direct link between the level of personal agency and the applied mode of action execution and action representation. High-level identification is associated with an appropriate level of identification for different types of actions, and high-level agents tend to have a greater efficacy in their everyday actions. However, since low-level identification is related to greater difficulties in adapting representations in response to action constraints, in low-level agents action flow is generally disrupted (for example, these individuals express stronger doubts about the satisfactory completion of an action). The poor performance and subsequent repetition of action may be related to the focus of attention on low-level gestural units of behavior rather than on the goal-related higher-level units that are normally used in action flow [39]. Thus, at one extreme is the low-level agent, who operates in the world primarily at the level of detail; at the other extreme is the high-level agent, who views his or her own actions in terms of causal effects, social meanings, and self-descriptive implications. Nevertheless, levels of personal agency do not represent a trait in the most common sense, but a tendency to adopt a behavior from within a content-defined class.

The AIT also proposes the existence of a consistent link between the manner by which people understand what they are doing and the way they understand themselves. High-level agents are inclined to extract intentions behind actions, which can provide meaningful depiction of the self. Low-level agents attach little significance to self-understanding, which can make them uncertain about what they are doing. Therefore, personal agency may be related to important aspects of the sense of self, such that a high-level profile allows subjects to extract abstract self-knowledge from actions and may give these individuals a coherent and stable understanding of themselves. On the other hand, a low-level of agency may produce a weaker and less coherent sense of self.

Recent studies have explored the relationship between specific clinical profiles and the level of agency. Specifically, dissociative states [40] and certain subclinical populations were shown to be sensitive to disruption of the level of agency [41]. *The behavior identification form (BIF)* [37] was designed to measure individual differences in action identification level across an array of routine actions. This approach was applied in a large sample in order to verify the relationship between "checker" behavior and low-level profile. The results showed that checking symptoms are related to the tendency to identify routine actions in terms of concrete mechanistic details. Ritualized actions are not connected to a representation of an accessible goal. In the absence of goal representations, gestural representations guide actions, which may lead people to be task- and performance-oriented rather than goal-oriented. In addition, in the carrying out of routine actions, acting that is guided by low-level representation may promote signals of inconsistency and error. Low-level agents may therefore prefer to control their actions according to situational cues, leading to the emergence of successive alternative meanings that necessitate updating the current action, which may engage these agents in new courses of action.

7.4.6
Lines of Research on the Disruption of Agency: ERPs and Personality

Recent studies used event-related potential (ERP) procedures to examine whether the awareness of being in control or not being in control of one's action correlates with specific brain activity [42]. Specifically, the external feedback content (true *vs* false) was manipulated in order to test the subjective response to these different types of feedback. The activity of the anterior cingulate cortex was associated with error detection between the subject's intention and the executed corresponding action [43, 44]. The discovery of specific neural correlates of behavior evaluation led to further research, in which a neural ERP response to errors, called ERN (error-related negativity), was identified [45]. Whereas initially the ERN was considered to reflect error detection, it was later suggested that it is involved in a more general evaluation of action plans [46] or conflict monitoring [43]. Furthermore, a related ERP effect, feedback-ERN (or FRN), was identified; it occurs in response to an external feedback, the amplitude of which is monotonically related to the degree of expectedness of the event, being larger for unexpected than for expected outcomes [47, 48]. The FRN is thought to represent the activity of a generic response monitoring system. It was also recorded during feedback indicating the incorrect performance of a time-production task and during distorted feedback [49-52].

An important question regarding the processing of external feedback (both erroneous and veridical) is whether subjective sensitivity to the external cues of reward *vs* punishment has an effect. Feedback perception and error-feedback may be directly related, and in some cases amplified, by an individual's personal features, i.e., his or her motivation and affective style, and by personal sensitivity to internal/external cues [53]. A prevalent view suggests that this subjective sensitivity corresponds to two general systems for orchestrating adaptive behavior [54, 55]. The first system functions to halt ongoing behavior while processing potential threat cues. It is referred to as the *behavior inhibition system* (BIS) [56]. A second system is believed to govern the engagement of action and has been referred to as the behavioral approach system [57] or the *behavioral activation system* (BAS) [58]. The BAS is conceptualized as a motivational system that is sensitive to signals of reward and non-punishment and is important for engaging behavior toward a reward. The BIS, conversely, inhibits behavior in response to stimuli that are novel, innately feared, and conditioned to be aversive [59].

The impact of external feedback on a person's behavior is determined, in addition to the contingent presence of external cues, by our personal sensitivity to these cues. Individual differences in action identification, explored by the BIF, and personal sensitivity to external feedback, analyzed by BIS/BAS measures [56], were considered as an explicative variable affecting cortical responses to error-feedback. Thus, a central question in neuropsychological research is how individual differences in feedback perception are manifested in motivation and personality, and how they can directly regulate cortical responses. In this context, the relationship between individual differences in action identification level and modulation of the cortical response to an external feedback has been examined.

Nevertheless, few studies have tried to connect brain response to the disruption of the sense of agency due to an external feedback, taking into account subjective attitude in terms of action representation (BIF) and behavioral predisposition (BIS/BAS) toward external context, and considering the impact of these factors on a veridical or a distorted external feedback processing. Based on these goals, it can be hypothesized that a FRN effect is found in response to an external feedback, with an increased effect in the case of a false condition, analogous to the previously described FRNs [49, 60]. In addition, a cortical system exists to process feedback response, i.e., the response to a congruent or incongruent system. Moreover, regarding the BIS/BAS effect on FRN, a significant sensitivity to feedback, in terms of a higher BIS, is expected especially for erroneous and negative feedback, due to the increased response to context of potential punishment, whereas a more proactive monitoring system should intervene to check for internal/external congruence in high-BAS subjects. Specifically, an attentional response to feedback should be marked by BAS; in BIF, however, a significant relationship between lower-level identities and feedback sensitivity is likely, in particular for false feedback since these identities are susceptible to a revised understanding of their behavior when the external cues signal a disrupted action or action-action effect relationship.

To summarize, the level of action identification has implications for understanding and controlling behavior. It is an independent dimension that may distinguish to what extent an individual has organized his or her action into abstract, meaningful categories that can operate to channel behavior into dispositional tendencies. Action identification theory holds that any action can be identified in many ways, ranging from low-level identities that specify how the action is performed, to high-level identities that signify why or with what effect the action is performed. People who identify action at a uniformly lower or higher level across many action domains, then, may be characterized in terms of their position within a broad personality dimension, and thus the level of personal agency. High-level agents think about their acts in encompassing terms that incorporate the motives and larger meanings of the action, whereas low-level agents think about their acts in terms of the details or means of action.

Research on behavior identification construct has examined the individual's overall competence in action and the degree to which his or her actions are organized by and reflected in the self-concept, and the implications of both. A significant relationship was established between low-level of action identification and increased FRN amplitude, particularly in response to false feedback. The level of identification most likely to be adopted by an actor is dictated by processes reflecting a trade-off between concerns for comprehensive action understanding and effective action maintenance. This suggests that the actor is always sensitive to contextual cues for higher levels of identification but moves to lower levels of identification if the action proves difficult to maintain with higher-level identities in mind. These processes have been documented empirically, as was their coordinated interplay in promoting a level of strict identification that matches the upper limits of the actor's capacity to perform the action. If this tendency has significance in terms of individual difference, then low-level agents should consider a wide assortment of actions to be more difficult and complex, and as a result they should be more sensitive than high-level agents to external cues or feedback.

Disruption of the sense of agency by erroneous feedback may directly implicate an immediate transposition from a high-level representation of action to a low-level representation, as suggested and predicted in an action identification model stating that the subject moves to lower levels of identification if the action proves difficult to maintain using higher-level identities [61]. In fact, previous research showed that people move to lower-level identities when an action is difficult, unfamiliar, or complex, when their performance of the act is disrupted [62], or when they are given failure feedback on their performance [63]. Empirical studies confirmed that disrupting people's action or otherwise inducing them to consider lower-level identities makes them susceptible to a revised understanding of their behavior, which, in turn, allows them to establish new courses of action [64].

More generally, a common denominator of the FRN effect and BIF measures could explain the distinction between personal planning and monitoring, on the one hand, and environmental control of actions, on the other. People are said to have either an internal or an external locus of control [65], to respond either in accord with personal standards or in response to the cues provided by others or by external feedback [66], and to behave consistently or inconsistently across time and situations. This distinction may involve internal action planning, which varies with differing levels of personal agency. Whether a person appears responsive to situational cues and constraints or instead maintains a personal plan of action despite the influence of these controls depends on his or her level of personal agency. High-level individuals are more aware of the overall planning of what they are doing and are less primed to accept signals that are provided by the context of the action. By contrast, low-level individuals, due to their lack of a high-level plan to integrate what they do, look to the information provided in the action's context to determine the significance of their behavior. Thus, changes in contextual cues are readily noticed and give emergent meaning to action. Overall, low-level agents might engage in impulsive as opposed to planned behavior, responding to salient cues in the particular situation, such as false-behavior feedback, whereas high-level agents might show greater internal control and stability in their representation across contextual variations.

References

1. Baron-Cohen S (1997) Mindblindness. An Essay on autism and theory of mind. MIT Press, Cambridge, MA
2. Rossetti Y, Pisella L (2002) Several "vision for action" system: a guide for dissociating and integratine dorsal and ventral functions. In: Prinz W, Hommel B (eds) Attention and Performance XIX: Common mechanisms in perception and action. Oxford University Press, Oxford, pp 62-119
3. De Vignemont F, Fourneret P (2004) The sense of agency: a philosophical and empirical review of the "Who" system. Conscious Cogn 13:1-19
4. Della Sala S, Marchetti C, Spinnler H (1991) Right-sided anarchic (alien) hand: a longitudinal study. Neuropsychologia 29:1113-1127
5. Riddoch JM (1990) Loss of visual imagery: a generation deficit. Cogn Neuropsychol 7:249-273
6. Weiskrantz L (1989) Blindsight. In: Boller F, Grafman J (eds) Handbook of neuropsychology. Elsevier, Amsterdam, pp 375-385
7. Paillard J (2005) Vectorial versus configural encoding of body space: a neural basis for a distinction between body schema and body image. In: De Preester H, Knockaert V (eds) Body imagery and body schema. Interdisciplinary perspectives on the body. John Benjamins, Amsterdam, pp 89-109
8. Rossetti Y, Gilles R, Farnè A, Rossetti A (2005) Implicit body representations in action. In: De Preester H, Knockaert V (eds) Body imagery and body schema. Interdisciplinary perspectives on the body. John Benjamins, Amsterdam, pp 111-125

9. Rossetti Y, Revonsuo A (2000) Beyond dissociation: interaction between dissociated implicit and explicit processing. John Benjamins, Amsterdam
10. Miltner WHR, Braun CH, Coles MGH (1997) Event-related brain potentials following incorrect feedback in a time-production task: evidence for a "generic" neural system for error detection. J Cognitive Neurosci 9:787-797
11. Milner AD, McIntosh, RD (2003) Reaching between obstacles in spatial neglect and visual extinction. Prog in Brain Res 144:213-226
12. Milner AD, McIntosh RD (2005) The neurological basis of visual neglect. Curr Opin Neurol 18:748-753
13. Jeannerod M, Rossetti Y (1993) Visuomotor coordination as a dissociable function: experimental and clinical evidence. In: Kennard C (ed) Visual perceptual defects. Baillère's Clinical Neurology, International Practise and Research. Ballière Tindall, London, pp 439-460
14. Vallar G (1997) Spatial frames of reference and somatosensory processing: a neuropsychological perspective. Philos T R Soc Lond B 352:1401-1409
15. Vecchi T, Bottini G (2006) Imagery and spatial cognition. Method, models and cognitive assessment. John Benjamins, Amsterdam
16. Bisiach E, Perani D, Vallar G, Bert A (1986) Unilateral neglect: personal and extrapersonal. Neuropsychologia 24:759-767
17. Vallar G, Bottini G, Paulesu E (2003) Neglect syndromes: the role of parietal cortex. In: Siegel AM, Andersen RA, Freund H-J, Spencer DD (eds) Advances in neurology, vol. 93. The parietal lobes. Lippincott Williams & Wilkins, Philadelphia, pp 293-319
18. Pia L, Neppi-Madona M, Ricci R, Berti A (2004) The anatomy of anosognosia for emiplegia: a meta analysis. Cortex 40:367-377
19. Blass DM, Hatanpaa KJ, Brandt J et al (2004) Dementia in hippocampal sclerosis resembles frontotemporal dementia more than Alzheimer disease. Neurology 63:492-497
20. Meltzoff AN, Moore MK (1995) Infant's understanding of people and things: from body imitation to folk psychology. In: Bermudez JL, Marcel A, Eilan N (eds) The body and the self. MIT Press, Cambridge, MA, pp 43-69
21. Gallese V (2001) The "Shared Manifold" hypothesis: from mirror neurons to empathy. J Consciousness Stud 8:33-50
22. Georgieff N, Jeannerod M (1998) Beyond consciousness of external reality: a "Who" system for consciousness of action and self-consciousness. Conscious Cogn 7:465-477
23. Frith CD, Done DJ (1989) Experiences of alien control in schizophrenia reflect a disorder in the central monitoring of action. Psychol Med 19:359-363
24. Jeannerod M, Farrer C, Franck N et al (2003) Recognition of action in normal and schizophrenic subjects. In: Kircher T, David A (eds) The self in neuroscience and psychiatry. Cambridge University Press, Cambridge, pp 380-406
25. Davies M, Coltheart M, Langdon R, Breen N (2001) Monothematic delusions: towards a two-factor account. In: On understanding and explaining schizophrenia, a special issue of Philosophy. Psychiatry and Psychology 8:133-158
26. Blakemore S-J, Sarfati Y, Bazin N, Decety J (2003) The detection of intentional contingencies in simple animations in patients with delusions of persecution. Psychol Med 33:1433-1441
27. Newcombe N (1989) Development of spatial perspective taking. In: Reese HW (ed) Advances in child development and behavior, vol. 22, pp 203-247
28. Meltzoff AN, Decety J (2003) What imitation tells us about social cognition: a rapprochement between developmental psychology and cognitive neuroscience. Philos T Roy Soc B 358:491-500
29. Gallagher S (2000) Philosophical conceptions of the self: implications for cognitive science. Trends Cogn Sci 4:14-21
30. Frith U, Frith CD (1999) Interacting minds. A biological basis. Science 286:1692-1695

31. Blakemore S-J, Rees G, Frith CD (1998) How do we predict the consequences of our actions? A functional imaging study. Neuropsychologia 36:521-529
32. David N, Gawronski A, Santos NS et al (2007) Dissociation between key processes of social cognition in autism: impaired mentalizing but intact sense of agency. J Autism Dev Disord 38:593-605
33. Daprati E, Wriessnegger S, Lacquaniti F (2007) Kinematic cues and recognition of self-generated actions. Exp Brain Res 177:31-44
34. Ramnani N, Miall RC (2004) A system in the human brain for predicting the actions of the others. Nat Neurosci 7:85-90
35. Synofzik M, Vosgerau G, Newen A (2008) Beyond the comparator model: a multifactorial two-step account of agency. Conscious Cogn 17:219-239
36. Cole ME, Frost RO, Heimberg RG, Rhéaume J (2003) "Not just right experiences": perfectionism, obsessive-compulsive features and general psychopathology. Behav Res Ther 41:681-700
37. Vallacher RR, Wegner DM (1989) Levels of personal agency: individual variation in action identification. J Pers Soc Psychol 57:660-671
38. Vallacher RR, Wegner DM (1985) A theory of action identification. Lawrence Erlbaum Associates, Hillsdale, New Jersey
39. Boyer P, Liénard P (2006) Why ritualized behavior? Prediction systems and action parsing in developmental, pathological and cultural rituals. Behav Brain Sci 29:1-56
40. Rufer M, Fricke S, Held D et al (2006) Dissociation and symptom dimensions of obsessive-compulsive disorder: a replication study. Eur Arch Psy Clin N 256:146-150
41. Belayachi S, Van der Linden M (in press) A French adaptation of the Behaviour Identification Form. In preparation
42. Balconi M, Crivelli D (2010) FRN and P300 ERP effect modulation in response to feedback sensitivity: the contribution of punishment-reward system (BIS/BAS) and behaviour Identification of action. Neurosci Res 66:162-172
43. Gehring WJ, Fencsik DE (2001) Functions of the medial frontal cortex in the processing of conflict and errors. J Neurosci 21:9430-9437
44. Holroyd CB, Krigolson OE, Baker R et al (2009) When is an error not a prediction error? An electrophysiological investigation. Cogn Affect Behav Ne 9:59-70
45. Falkenstein M, Hohnsbein J, Hoormann J, Blanke L (1990) Effects of errors in choice reaction tasks on the ERP under focused and divided attention. In: Brunia CHM, Gaillard AWK, Kok A (eds) Psychophysiological brain research. Tilburg University Press, Tilburg, Netherlands, pp 192-195
46. Luu P, Flaisch T, Tucker DM (2000) Medial frontal cortex in action monitoring. J Neurosci 20:464-469
47. Balconi M, Crivelli D (2009) Spatial and temporal cognition for the sense of agency: neuropsychological evidences. Cogn Process 10:182-184
48. Heldmann M, Russeler J, Munte T (2008) Internal and external information in error processing. BMC Neurosci 9:1-8
49. Miltner WHR, Braun CH, Coles MGH (1997) Event-related brain potentials following incorrect feedback in a time-estimation task: evidence for a "generic" neural system for error detection. J Cognitive Neurosci 9:788-798
50. Ehlis A-C, Herrmann MJ, Bernhard A, Fallgatter AJ (2005) Monitoring of internal and external error signals. J Psychophysiol 19:263-269
51. Müller SV, Möller J, Rodriguez-Fornells A, Münte TF (2005) Brain potentials related to self-generated and external information used for performance monitoring. Clin Neurophysiol 116:63-74

52. Vocat R, Pourtois G, Vuilleumier P (2008) Unavoidable errors: a spatio-temporal analysis of time-course and neural sources of evoked potentials associated with error processing in a speeded task. Neuropsychologia 46:2545-2555
53. Dikman ZV, Allen JJB (2000) Error monitoring during reward and avoidance learning in high- and low-socialized individuals. Psychophysiology 37:43-54
54. Gray JA (1981) A critique of Eysenck's theory of personality. In: Eysenck HJ (ed) A model for personality. Springer, Berlin, pp 246-277
55. Carver CS, White TL (1994) Behavioral inhibition, behavioral activation, and affective responses to impending reward and punishment: the BIS/BAS scales. J Pers Soc Psychol 67:319-333
56. Gray JA (1990) Brain systems that mediate both emotion and cognition. Cognition Emotion 4:269-288
57. Gray JA (1982) The neuropsychology of anxiety: an inquiry into the functions of the septo-hippocampal system. Oxford University Press, New York
58. Fowles DC (1980) The three arousal model: implications of Gray's two-factor learning theory for heart rate, electrodermal activity, and psychopathy. Psychophysiology 17:87-104
59. Boksem MAS, Tops M, Wester AE et al (2006) Error-related ERP components and individual differences in punishment and reward sensitivity. Brain Res 1101:92-101
60. Holroyd CB, Coles MGH (2002) The neural basis of human error processing: reinforcement learning, dopamine, and the error-related negativity. Psychol Rev 109:679-709
61. Vallacher RR, Nowak A, Markus J, Strauss J (1998) Dynamics in the coordination of mind and action. In: Kofta M, Weary G, Sedek G (eds) Personal control in action: cognitive and motivational mechanisms. Plenum, New York, pp 27-59
62. Wegner DM, Vallacher RR, Macomber G et al (1984) The emergence of action. J Pers Soc Psychol 46:269-279
63. Vallacher RR, Wegner DM, Frederick J (1987) The presentation of self through action identification. Soc Cognition 5:301-322
64. Wegner DM, Vallacher RR, Kiersted GW, Dizadji DM (1986) Action identification in the emergence of social behavior. Soc Cognition 4:18-38
65. Rotter JB (1966) Generalized expectancies for internal versus external control of reinforcement. Psychol Monogr-Gen A 80:1-28
66. Snyder M, Campbell BH (1982) Self-monitoring: the self in action. In: Suls J (ed) Psychological perspectives on the self. Erlbaum, Hillsdale, New Jersey, pp 185-207

Disturbances of the Sense of Agency in Schizophrenia

M. Synofzik, M. Voss

8.1 Introduction

The sense of agency and its neurocognitive underpinnings have been the subjects of increasing attention over the last several years, but their detailed mechanisms remain controversial. An excellent opportunity to investigate both the basic neurocognitive mechanisms of self-agency attribution and their pathological dysfunctions is to study abnormalities of the sense of agency in neurological or psychiatric patients. In particular, disturbances of agency processing in schizophrenia patients with delusions of influence might reveal specific central mechanisms for the self-attribution of agency, which can be specifically impaired. Patients with delusions of influence feel that someone else is guiding and executing their actions, even if the action is actually completely caused by themselves.

Here we review recent findings on delusions of influence in schizophrenia, demonstrating that these findings suggest a new general framework of agency processing. This framework integrates the influential *comparator model* of agency with the latter's seemingly contradictory findings. Moreover, and importantly, the new approach extends the comparator model to account for the large variety of available internal and external (e.g., sensory, cognitive, and contextual) cues, which likewise contribute to agency processing and may play a particular role in delusions of influence in schizophrenia. This framework may explain not only delusions of influence, but also a larger variety of disturbances in agency attribution in psychotic patients.

M. Synofzik (✉)
Department of Neurodegeneration, Center of Neurology & Hertie–Institute for Clinical Brain Resarch, Tübingen, Germany

Neuropsychology of the Sense of Agency. Michela Balconi (Ed.)
© Springer-Verlag Italia 2010

8.2
The Comparator Model and Its Explanatory Limitations

Already in 1950, two studies [1, 2] suggested a simple and attractive computational mechanism, also referred to as the comparator model, to explain how the nervous system might deal with the self-world distinction when attributing agency to sensory events. According to this account, the sensory consequences of one's actions can be predicted based on an internal *efference copy* of the motor command [2] or an internal *corollary discharge* [1]. In order to isolate externally produced sensory stimuli, afferent stimuli are constantly compared with these internal predictions. In case of a match, the afference is interpreted as a result of self-action. In case of a mismatch, the difference corresponds to an externally caused event.

In fact, both psychophysical and electrophysiological studies show that the constant comparison between internal predictions and external information ensures that we correctly attribute self-produced sensations to our own agency rather than to external causal forces [3-7]. This mechanism allows us, for instance, to cancel out potentially disturbing self-produced sensory events, e.g., self-produced tactile stimulation [3] or visual flow due to our own pursuit eye movements, thereby guaranteeing the perception of a stable world despite self-motion [8,9].

Following the suggestions of Feinberg [10], this model was subsequently applied by Frith and colleagues to study cases of pathological misattribution of agency [11,12]. It was considered that failures in the of self-attribution to the sensory consequences of one's actions might result from an impairment in generating adequate internal predictions and/or in comparing internal predictions with the actual sensory afference (Fig. 8.1). According to this idea, patients would be expected to attribute any deviant sensory information (which is no longer compensated by internal prediction) to external sources rather than to themselves–as is the case in delusions of influence.

Indirect evidence for such prediction deficits in schizophrenia patients comes from two recent behavioral studies. The first used a novel force-matching task [13, 14]. In this experiment, subjects experienced a force applied to their finger by a torque motor and were then required to match the perceived force by actively pushing on the finger using their other hand. Due to an attenuation of predictable sensory input, healthy subjects reliably under-estimated the force they were applying and thus produced a much larger active force than was experienced passively [13]. Conversely, schizophrenia patients showed significantly less attenuation than age-matched healthy control subjects, suggesting an impaired sensory predictive mechanism in schizophrenia [14].

The second study investigated the ability to cancel out self-induced retinal image motion resulting from one's own smooth-pursuit eye movements [9]. During performance of smooth pursuit, the images of a stationary environment inevitably slip over the retina with a velocity equivalent to that of the eye's rotation. If we relied on retinal information only, we would misattribute image motion to the environment rather than to ourselves and thus misperceive the world as moving. This interpreta-

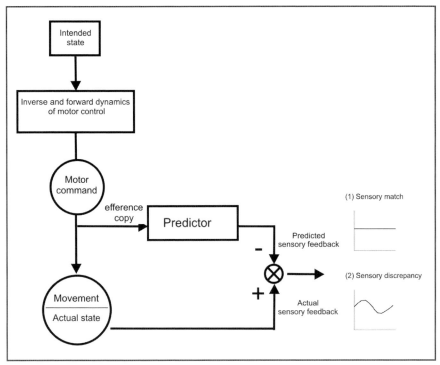

Fig. 8.1 The neurocognitive comparator mechanism underlying the sense of agency. On the basis of a given motor command, the system predicts the outcome of one's own behavior. By comparing the predicted sensory consequences with the actual sensory afference, self-produced sensory information can be distinguished from externally caused events. In case of a match, the afference is interpreted as a result of one's own action. In case of a mismatch, the difference is registered as externally caused [1, 2]. An impairment in predicting the sensory consequences of one's own movements would lead to a mismatch at the comparator and, consequently, to false registration of self-produced sensory events as externally caused

tion is avoided by comparing the actual image slip with the amount of image motion predicted on the basis of an efference copy of the eye-movement motor command. If both signals match, the retinal image slip is interpreted as being self produced and cancelled out; if they do not match, a residual motion difference is perceived which must be attributed to the external world [8]. Consequently, if internal predictions were imprecise in schizophrenia patients with delusions of influence, these patients should perceive a greater amount of residual motion. Lindner and colleagues demonstrated that schizophrenia patients with delusions of influence were impaired in perceptually compensating for smooth-pursuit-induced image motion, thus again indicating an impaired sensory predictive mechanism in schizophrenia [9].

Most of the other empirical findings which were commonly taken as support for a role of the comparator model in schizophrenia demonstrated an exaggerated *self-attribution* of sensory events (e.g., "hyper-associations" between an action and its effect [15]) in schizophrenia patients. For example, several studies reported that in tasks with spatial or temporal distortion of visual feedback of one's own hand movements, schizophrenia patients expressed a stronger tendency to attribute what they saw to their own actions [15-19]. These findings cannot be directly explained by a deficit of the comparator mechanism: due to the lack of adequate internal predictions that would compensate for the self-produced sensory events, patients should show an exaggerated *external* attribution of sensory events. Similarly, the psychopathology of delusions of influence predominantly reflects an under-attribution (not an over-attribution) of self-produced sensory information.

In addition, it remains unclear how a deficit in internal predictions could be sufficient to cause the misattribution of actions observed in schizophrenia. In order to transit from abnormal experience to delusional belief, unusual belief-formation processing must be postulated. Only such an idiosyncrasy in the belief-formation process could explain why schizophrenia subjects (1) do not accept an alien experience as a strange experience (as, for example, healthy subjects or most neurological patients with alien motor phenomena would do) but (2) devise a delusional agency hypothesis about this experience and (3) maintain it despite different stored encyclopedic knowledge about their behavior and despite the testimony of others [20]. Thus, an explanation of delusions of agency also needs to account for the abnormalities in the belief-formation system with respect to a person's method of action rationalization and self-theorizing. A deficit in internal predictions per se also cannot provide any explanation for the semantics of the delusional belief: Why is it that agency attribution fails only in certain semantic contexts that are often highly specific to the history of the delusional individual? And why does it have its specific semantic content (e.g., an action is caused by a stranger or by God)? To explain this, one would have to integrate information from a person's broader belief system and narrative self-structure. In other words, if the thesis of impaired internal predictions as underlying delusions of influence is correct, it can only explain why schizophrenia subjects have an abnormal experience of their own action. Yet this is not enough to explain why they devise delusional beliefs with certain content [21].

8.3
Feeling of Agency *vs* Judgement of Agency

One of the reasons for the discrepancy between experimental findings and predictions based on the comparator model seems to be rather simple. The sense of agency might not function as a unitary processing module (as it appears phenomenally), but in fact represent a complex supramodal phenomenon of largely heterogeneous functional and representational levels, with different agency cues receiving a different weighting on each level [21, 22]. Previous experimental approaches might simply operate on levels of agency processing different than those of the comparator model, whereby the experimental findings might not directly reflect internal predictions and/or the comparator output.

The sense of agency comprises at least two different levels of agency registration [21] (Fig. 8.2): on the level of agency attribution–which has been tested by most self-recognition studies–subjects have to make explicit *judgments* about the agent of an action. This level integrates many complex cognitive cues, e.g., prior expectations about the task, background beliefs, and context estimations, but it does not directly reflect the immediate *feeling* of agency. It is this default level of agency that is most prevalent in our everyday life: when we grasp, type, or walk, our sensorimotor system implicitly registers these sensory consequences as self-caused and they are withheld from further, demanding processing and in particular from further rationaliza-

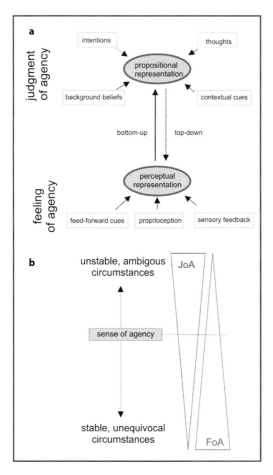

Fig. 8.2 The sense of agency is construed by different agency levels and by a context-dependent, flexible weighting of agency cues. (**a**) The sense of agency comprises at least two different levels of agency registration. The basic non-conceptual feeling of agency is produced by a gradual and highly plastic subpersonal weighting process of different action-related perceptual and motor cues (feeling of agency). This pre-conceptual core is further processed by conceptual capacities and attitudes (e.g., beliefs, desires) to form an attribution of agency (judgement of agency). Both levels closely interact: the feeling of agency can determine our judgement of agency and, in turn, cues from the judgement level can largely influence our feeling of agency. Thus, the sense of agency is a dialectic combination of bottom-up and top-down processes. (**b**) The extent to which the feeling of agency and the judgement of agency, respectively, contribute to the overall sense of agency depends on the context and task requirements. In stable, unequivocal situations, our sense of agency is fully determined by the feeling of agency. In these situations, our sensorimotor system implicitly registers the sensory consequences of its actions as self-caused and they are withheld from further demanding processing and, in particular, further propositional modules. In more ambiguous situations, the system resorts more to contextual cues, background beliefs, and other propositional modules to infer agency of sensory events. (Modified from [21])

tion modules. This basic representation is commonly thought to depend mainly on the coherence of motor and sensory cues related directly to the action itself, and especially on internal predictions and/or the comparator output [21].

Thus, it might well be that at least some findings of over-attribution/hyper-association of sensory events in schizophrenia patients result from the fact that in the respective studies explicit agency judgements were tested (e.g., by asking "was it you or not you?") and that for some reason, yet to be explained, schizophrenia patients tend to over-attribute sensory events to their own agency despite a mismatch at the underlying comparator. Likewise, it is possible that an impairment on the level of feeling of agency is complemented by an impairment on the level of judgement of agency, the combination of both resulting in a delusional agency belief.

8.4
Optimal Cue Integration as the Basis of the Sense of Agency

One option to resolve the seeming discrepancy between empirical findings in self-recognition tasks and model predictions is by explaining over-attributions of sensory events in schizophrenia patients as the result of an over-reliance on external sensory action events, which is, in turn, due to an under-reliance on internal action cues. And, in fact, according to the comparator model, schizophrenia patients should be predisposed to under-rely on their internal predictions as these predictions are postulated to be inadequate. This explanation extends the comparator model by a framework of *optimal cue integration*, according to which no single information signal is powerful enough to convey an adequate representation of a certain perceptual entity under all everyday conditions. Instead, depending on the availability and reliability of a certain information cue (here, e.g., internal predictions and external sensory information), different combination and integration strategies should be used to frame the weighting of sensory and motor signals. Thus, the type of optimal cue integration might not only allow robust perception of the world [24, 25] and efficient sensorimotor learning [26], it could also provide the basis for subjects' robust, and at the same time flexible, agency experience in variable contexts: the sense of agency constantly reflects the relative reliability of the respective agency cues in a given situation [22].

Within this integration, intrinsic efferent signals such as internal predictions probably serve as the most reliable and robust agency cues, as they usually provide the fastest and least noisy information about one's own actions [27]. However, other cues might outweigh or even replace these efferent signals to install a basic registration of agency. For example, it was shown that even for involuntary movements, i.e., movements for which no internal predictions are issued, an implicit registration of agency can be installed if prior action cues (here: primes) are present [28]. This finding can be well explained within the framework of optimal cue integration: involuntary movements might be particularly susceptible to prior action cues as these cues should receive particularly more weight when one of the most robust cues–namely, efference-copy-based predictions–is absent. Thus, under normal conditions, the registration of being the initiator of one's own actions seems to arise from a dynamic interplay between prior expectations, internal predictions about the sensory consequences of one's actions, sensory information, and post-hoc beliefs. These cues are not mutually exclusive, but used in combination according to their respective reliability to establish the most robust agency representation in a given situation.

8.5
Altered Cue Integration as the Basis of Delusions of Influence

The approach of optimal cue integration might provide a common basis for the various misattributions of agency in schizophrenia patients. In schizophrenia, and even more so in an acute psychotic state, internal predictions about the sensory consequences of one's actions could be frequently imprecise and non-reliable. Patients should therefore be prompted to rely more on (seemingly more reliable) alternative cues about self-action, such as vision, auditory input, prior expectations, or post-hoc thoughts. The stronger weighting of these alternative cues could help patients to avoid a misattribution of agency for self-produced sensory events that was due to imprecise internal predictions. However, as a consequence of giving up the usually most robust and reliable internal action information source, i.e., internal predictions, the sense of agency in psychotic patients is at constant risk of being misled by ad hoc events, invading beliefs, and confusing emotions and evaluations. Different agency judgement errors may result: patients might over-attribute external events to their own agency whenever these more strongly weighted alternative agency cues are not veridical and misleading, as is the case in delusions of reference (also referred to as "megalomania"). Conversely, if alternative cues are temporarily not attended or unavailable, patients might fail to attribute self-produced sensory events to their own agency and instead assume external causal forces (as is the case in delusions of influence). A context-dependent weighted integration of imprecise internal predictions and alternative agency cues may therefore reflect the basis of agency attribution errors in both directions: over-attribution, as in delusions of reference/megalomania, and under-attribution, as in delusions of influence.

8.5.1
Intentional Binding: Impaired Predictions and Excessive Linkage of External Sensory Events

Two recent studies provided the first empirical evidence supporting this notion. Voss and colleagues [29] used an elaborate version of the "intentional binding" paradigm [30] to study the relation between predictive and sensory effect-driven ("postdictive") mechanisms of agency processing in schizophrenia patients. These authors measured the perceived onset time of a self-initiated key press that caused an external sensory effect (a tone) with high probability (75%), with chance probability

(50%), or with 0% probability (baseline condition). The perceived action-onset times in these conditions were compared. In the high-probability condition, healthy controls showed a shift of the perceived time of action onset towards the tone compared to the baseline condition. Crucially, this shift also occurred on occasional trials in which the tone was omitted. In contrast, schizophrenia patients did not show any shift of perceived action-onset time in the absence of the tone.

This finding was taken as evidence for a deficit in forming internal predictions about the sensory effects of one's actions: schizophrenia patients seem to be unable to use previously acquired experience of a certain action-effect-linkage to predict the sensory effect when it is occasionally omitted. Interestingly, the degree of this deficit correlated with different aspects of the patients' psychopathology; inter alia, with specific positive psychotic symptoms: the stronger delusions and hallucinatory behavior, the smaller the shifts in perceived action onset. When the tone actually occurred, however, patients showed a larger shift of perceived action onset towards the tone than healthy controls. This finding indicates that patients' temporal binding between action and effect is largely driven by the actual presence of a sensory event. Thus, rather than using predictive mechanisms, schizophrenia patients seem to solely rely on external information about the action and to infer agency by a "postdictive" process.

8.5.2
Perception of Hand Movements: Imprecise Predictions Prompting an Over-reliance on External Action Cues

From the afore-mentioned study, it remains an open question whether impaired prediction and exaggerated reliance on external sensory information in schizophrenia patients work in parallel, thus reflecting two different processes, or whether it is *by* the impairment in predictions that patients over-rely on external information, thus reflecting a single causal process. The latter notion would provide more direct evidence for optimal cue integration underlying the sense of agency. Synofzik and colleagues studied the relationship between impaired internal predictions and over-reliance on external sensory information in two action perception experiments [23]

Subjects performed pointing movements in a virtual-reality setup in which the visual consequences of one's own movements could be rotated with respect to the actual movement. In the first experiment, subjects had to detect clockwise and counter-clockwise rotations of the visual feedback with respect to their actual movements. The patient group revealed higher thresholds for detecting experimental feedback rotations and, importantly, the size of these thresholds correlated positively and very selectively with patients' delusions of influence: the larger the impairment in detecting visual feedback distortions about their own movements, the stronger were the delusions of influence that patients experienced. This finding does not just confirm and specify previous findings of larger detection thresholds in the perception of self-action in schizophrenia patients, it also provides a specific explanation: subjects might have relied on the fact that the cursor was a visual representation of their own fingertip and used this (misleading) visual ownership information as an "external agency cue". Their perceptual system may have relied more strongly on this external action-related information in order to receive a more reliable account of one's own actions, as the system's own internal predictions are imprecise. The second experiment served to further assess both the accuracy of internal predictions and the weighing of internal and external action information. Here, subjects were required to estimate their direction of pointing visually in the absence of any visual feedback and again in the presence of visual feedback that was constantly rotated by 30°. The accuracy of internal predictions was assessed based on the trial-by-trial variability of subjects' visual self-action estimates in those trials without any visual feedback. The standard deviation of self-action estimates was significantly higher in schizophrenia patients and, importantly, correlated selectively with delusions of influence. In other words, the more imprecise the estimates of internal predictions, the stronger were the delusions of influence that patients experienced. Moreover, a significant correlation was found between the pooled standard deviation of subjects' self-action estimates and the detection thresholds determined in the first experiment. This observation suggests that imprecise internal predictions indeed underlie patients' impairment in perceiving their own actions as they prompt patients to over-rely on (potentially misleading) visual feedback. Indirect support for this assumption stems from subjects' level of adaptation in those trials involving (constantly rotated) visual feedback: patients adapted their self-action estimates almost twice as much as controls in the direction of the rotated visual feedback. The degree of this adaptation of self-action estimates relative to the absolute visual feedback rotation of 30° reflects a relative weighing of internal and external self-action information according to the respective relative reliability. The reliability of patients' internal predictions (estimated by the inverse of the squared pooled standard deviation of the perceptual self-action estimates in trials without feedback) was about half that of healthy controls (0.005 *vs* 0.01, 1/degree^2, respectively). Accordingly, schizophrenia patients should give less weight to internal predictions and increase the relative weight of alternative (visual) cues about self-action. Corresponding with this prediction, the average degree of adaptation to the visual feedback rotation was about two times larger in schizophrenia patients (63%; 18.9°) than in controls (37%; 11.0°).

8.6
Conclusions

The results of the two studies discussed in the previous section suggest that psychotic patients cannot resort to a reliable predictive structuring link between the self and external events. Such a link is, however, needed to reliably learn the connection between actions and effects, and thus to understand the relation between the self and the world. If this link cannot be established, the relation between the self and the world becomes blurred and confusing.

Under most circumstances, a stronger weighting of external cues will help patients to achieve more perceptual certainty about their own actions and to compensate for imprecise internal predictions. However, as a consequence of giving up the usually highly robust and reliable internal action information source, namely, internal predictions, the sense of agency in psychotic patients is at constant risk of being misled by ad hoc events, invading beliefs, and confusing emotions and evaluations. In other words, the sense of agency in psychotic patients is built on a fluctuating, unreliable basis and the lack of a coherent integration of the different, partly contradicting agency cues may lead to delusional explanations of subjective experiences related to one's own actions, such as occurs in delusions of control or of reference. The framework of optimal cue integration might thus be able to integrate the popular comparator model of agency, to overcome its explanatory limitations and to provide a unified account for the various delusions of agency in psychotic patients.

References

1. Sperry R (1950) Neural basis of the spontaneous optokinetic response produced by visual inversion. J Comp Physiol Psychol 43:482-489
2. von Holst E, Mittelstaedt H (1950) Das Reafferenzprinzip. Naturwissenschaften 37:464-476
3. Blakemore SJ, Frith CD, Wolpert DM (1999) Spatio-temporal prediction modulates the perception of self-produced stimuli. J Cogn Neurosci 11:551-559
4. Bays PM, Wolpert DM, Flanagan JR (2005) Perception of the consequences of self-action is temporally tuned and event driven. Curr Biol 15:1125-1128
5. Synofzik M, Thier P, Lindner A (2006) Internalizing agency of self-action: perception of one's own hand movements depends on an adaptable prediction about the sensory action outcome. J Neurophysiol 96:1592-1601
6. Voss M, Ingram JN, Haggard P, Wolpert DM (2006) Sensorimotor attenuation by central motor command signals in the absence of movement. Nat Neurosci 9:26-27
7. Crapse TB, Sommer MA (2008) Corollary discharge across the animal kingdom. Nat Rev Neurosci 9:587-600
8. Haarmeier T, Bunjes F, Lindner A et al (2001) Optimizing visual motion perception during eye movements. Neuron 32:527-535
9. Lindner A, Thier P, Kircher TT et al (2005) Disorders of agency in schizophrenia correlate with an inability to compensate for the sensory consequences of actions. Curr Biol 15:1119-1124
10. Feinberg I (1978) Efference copy and corollary discharge: implications for thinking and its disorders. Schizophr Bull 4:636-640
11. Frith C (1992) The cognitive neuropsychology of schizophrenia. Erlbaum, Hillsdale, NJ, USA
12. Frith CD, Blakemore S, Wolpert DM (2000) Explaining the symptoms of schizophrenia: abnormalities in the awareness of action. Brain Res Brain Res Rev 31:357-363
13. Shergill SS, Bays PM, Frith CD, Wolpert DM (2003) Two eyes for an eye: the neuroscience of force escalation. Science 301:187
14. Shergill SS, Samson G, Bays PM et al (2005) Evidence for sensory prediction deficits in schizophrenia. Am J Psychiatry 62:2384-2386
15. Haggard P, Martin F, Taylor-Clarke M et al (2003) Awareness of action in schizophrenia. Neuroreport 14:1081-1085

16. Daprati E, Franck N, Georgieff N et al (1997) Looking for the agent: an investigation into consciousness of action and self-consciousness in schizophrenic patients. Cognition 65:71-86
17. Fourneret P, Franck N, Slachevsky A, Jeannerod M (2001) Self-monitoring in schizophrenia revisited. Neuroreport 12:1203-1208
18. Franck N, Farrer C, Georgieff N et al (2001) Defective recognition of one's own actions in patients with schizophrenia. Am J Psychiatry 158:454-459
19. Knoblich G, Stottmeister F, Kircher T (2004) Self-monitoring in patients with schizophrenia. Psychol Med 34:1561-1569
20. Davies M, Coltheart M, Langdon R, Breen N (2001) Monothematic delusions: towards a two-factor account. Philos Psychiatr Psycholog 8:133-158
21. Synofzik M, Vosgerau G, Newen A (2008) Beyond the comparator model: a multifactorial two-step account of agency. Conscious Cogn 17:219-239
22. Synofzik M, Vosgerau G, Lindner A (2009) Me or not me–an optimal integration of agency cues? Conscious Cogn 18:1065-1068
23. Synofzik M, Thier P, Leube DT et al (2010) Misattributions of agency in schizophrenia are based on imprecise predictions about the sensory consequences of one's actions. Brain 133:262-271
24. Ernst MO, Banks MS (2002) Humans integrate visual and haptic information in a statistically optimal fashion. Nature 415:429-433
25. Ernst MO, Bulthoff HH (2004) Merging the senses into a robust percept. Trends Cogn Sci 8:162-169
26. Kording KP, Wolpert DM (2004) Bayesian integration in sensorimotor learning. Nature 427:244-247
27. Wolpert DM, Flanagan JR (2001) Motor prediction. Curr Biol 11:R729-732
28. Moore JW, Wegner DM, Haggard P (2009) Modulating the sense of agency with external cues. Conscious Cogn 18:1056-1064
29. Voss M, Moore J, Hauser M et al (2010) Altered awareness of action in Schizophrenia: a specific deficit in predicting action consequences. Brain; in press
30. Haggard P, Clark S, Kalogeras J (2002) Voluntary action and conscious awareness. Nat Neurosci 5:382-385

Looking for Outcomes: The Experience of Control and Sense of Agency in Obsessive-compulsive Behaviors

S. Belayachi, M. Van der Linden

9.1 Introduction

Obsessive-compulsive disorder (OCD) is, as its name implies, characterized by obsessions (i.e., recurrent thoughts or images, particularly ones that cause distress) as well as compulsions (i.e., urges to perform mental or physical acts repeatedly), both of which significantly impair everyday functioning [1]. Obsessions are considered to be recurrent distressing impressions that something is wrong with an action or with a situation, such as an error or an imminent danger [2–4]. Compulsions are generally conceptualized as aiming to prevent feared harmful events and are thus associated with an increased sense of responsibility [5, 6]. However, they may also be motivated by particular sensory experiences concerning actions, such as feelings of incompleteness, that trigger the need to adjust them, rather than the avoidance of potential harm [7].

Both the sense of responsibility and the feeling of incompleteness can be viewed as experiences of actions that are obviously in conflict with the actual action context. Specifically, compulsions can be seen as behavioral responses to recurrent feelings of dissatisfaction regarding an intended achievement. Moreover, OCD features have been consistently connected to deficits affecting action processing, such as action planning [8, 9] and action monitoring [10, 11].

Research on the sense of agency has highlighted the importance of action specification (e.g., outcome anticipation) and action monitoring (e.g., assessing the degree of concordance between anticipated and actual outcomes) in the subjective understanding of "what one is doing" and "what one is causing" [12–17]. Hence, a dysfunction affecting these components of action processing could lead to an incon-

S. Belayachi (✉)
Department of Cognitive Science, University of Liège, Liège, Belgium

sistent appraisal of one's actions and of the surrounding environment. More importantly, recent models of sense of agency suggest that there are various levels of action specification and monitoring (e.g., sensorimotor, perceptual, and conceptual levels); each level may contribute to a specific component of the sense of agency [14, 18]. Therefore, the nature of the impairment and the level at which the dysfunction applies may underlie specific patterns of a defective sense of agency.

In this chapter, we review the phenomenology of OCD to understand how its heterogeneity can be depicted as an outcome processing issue and how it can be differentially affected across distinct OCD profiles.

9.2
The Clinical Features and Phenomenology of OCD

Obsessive-compulsive disorder is commonly viewed as a set of maladaptive habits and ways of thinking, of which obsessions and compulsions are the most prominent symptoms. As in other psychopathological states, OCD symptoms are present to some degree in most people; their frequency and their impact on everyday functioning may distinguish non-clinical from clinical cases of OCD [19]. Obsessions involve intrusive thoughts, images, or impulses that cause significant distress. Common obsessions include preoccupation with contamination, concerns about potential threatening outcomes, fear of harming oneself or others, repeated doubts about self-action, and preoccupation with action satisfaction. Over 90% of patients with OCD report performing compulsive behaviors to reduce the distress associated with obsessions [20]. Compulsions refer to the need to perform mental or physical acts in a repeated or stereotyped way. Repetitive checking (e.g., checking locks, lights, and appliances) and repetitive washing (e.g., hand washing, house cleaning) are the most common compulsions [21, 22].

An important feature of OCD is that some behaviors and activities that are performed automatically by most people (e.g., washing one's hands, locking a door, tidying up clothes) are related to a dysfunctional experience of action. Examples include an inflated sense of responsibility for the occurrence or avoidance of bad outcomes, beliefs that one's thoughts can have direct negative consequences for the external world, beliefs that errors can have harmful consequences, an exaggeration of the probability and severity of potential harm, inconsistent impressions of failure or feelings of imperfection, feelings that actions or intentions have been incompletely executed, and an undermined sense of goal satisfaction. These inconsistent feelings and beliefs form strong motivational features concerning the occurrence of OCD symptoms. They can be classified according to the extent to which OCD symptoms are predominantly characterized by feelings of responsibility or of incompleteness.

Responsibility in OCD refers to the "belief that one has power which is pivotal to bring about or prevent subjectively crucial negative outcomes" ([23], p. 111). This OCD dimension seems to be dominated by cognitive phenomena, including dysfunctional beliefs, biased inferences and judgments, misinterpretation of intrusive thoughts. One important feature of this phenomenon is the premonitory aspect of negative outcomes. Indeed, OCD individuals experience the content of certain intrusive thoughts as an indication of a future negative consequence of their action or inaction. They may also voluntarily try to foresee a wide range of negative consequences of their actions. In addition, the processing of anticipated negative outcomes is influenced by dysfunctional beliefs, such as the idea that doing nothing about potentially upcoming bad events is equivalent to causing disastrous outcomes. Overall, these features lead OCD individuals to experience feelings of responsibility and guilt and compel them to undertake preventive actions (i.e., avoidance strategies).

On the other hand, incompleteness and "not just right" feelings are dominated by sensory phenomena and are described by patients as impressions of failure or feelings of imperfection. Such feelings can lead to an inability to achieve a sense of "task completion" or "closure" regarding actions (e.g., locking the door) or perceptions (e.g., objects on a table). This sense of dissatisfaction may cause people to experience inconsistent feelings that "actions or intentions have been incompletely achieved" ([24], p. 80). It may also lead them to feel only a weak sense of goal satisfaction. Repeated compulsions may then be motivated by the need to alleviate feelings of incompleteness or to feel "just right" [25].

Although these two distinct motivational core features may be more fundamental than mere symptom clusters, their prevalence seems to vary across the different OCD subtypes. Indeed, harm avoidance may particularly be reflected in washing and obsessing symptoms [26]. Incompleteness, on the other hand, may be especially associated with checking [26, 27] and ordering [26, 28]. Furthermore, distinct neural correlates have been associated with these different OCD subtypes. For example, checking symptoms, which are frequently related to incompleteness, may be connected to increased activity in brain regions involved with motor processing, such as dorsolateral prefrontal regions, putamen/globus pallidus, and brainstem nuclei [29–31]. Washing symptoms, which are related to harm avoidance, have been found to be associated with increased activity in brain regions that process emotional aspects of information (e.g., orbitofrontal regions [29, 31]).

Overactive performance monitoring, as reflected by the anterior cingulate cortex hyperactivity and electrophysiological abnormalities observed in OCD, has consistently been connected to incompleteness features [11, 32-34]. It is assumed to reflect excessive error detection, caused by an impaired comparison between actual and expected responses [35, 36]. Indeed, an internal comparator mechanism may compare the internal representation of action and the resulting action. If a conflict is detected, the system triggers a signal and adjustment behavior is activated. It has been suggested that a hyperactive error signal in OCD arises from a dysfunction in a comparator mechanism, which then erroneously detects a mismatch between representations of the actual and the intended response.

On the other hand, responsibility features seem to be associated with increased activity in brain regions that process the emotional aspects of action, such as the orbitofrontal cortex (OFC). The OFC may play an important role in action control and guidance of behaviors, through outcome representations and particularly by anticipating the affective impact of outcomes. This form of anticipation plays an important role in decision-making [37]. The representation of anticipated outcomes may depend on how the OFC generates possible alternative outcomes of one's action. Moreover, lateral areas of the OFC may underlie the anticipation of potential negative outcomes, while areas from the ventral and medial prefrontal cortex may be specifically involved in representing the impact of outcomes with a positive valence [38, 39]. Increased activity in the lateral OFC in OCD individuals has been consistently related to their concerns with potential future negative outcomes [40].

Taken together, the above-mentioned studies suggest that OCD symptoms are related either to an impaired ability to re-integrate generated outcomes as being consistent with intended outcomes, or to an increased processing of outcomes, particularly of their emotional value at an early stage of action specification. From this perspective, compulsions can be conceptualized as behavioral strategies aimed at either generating outcomes that match the intended ones, as is the case in incompleteness phenomena, or avoiding the occurrence of potential negative consequences.

To sum up, people with OCD may experience a sense of responsibility for threatening events whose occurrence are not related to their actions. Yet, these individuals feel compelled to deploy compulsions in order to counter bad outcomes. Other OCD individuals, however, may perform compulsive behaviors because they have inconsistent feelings that an action has not been satisfactorily completed or that their goals and intentions have not been achieved. Compulsive behaviors related to incompleteness phenomena have the purpose of generating outcomes that will provide a sense of task completion. Furthermore, the phenomenal state preceding compulsions can be related either to "feelings of undesired end-state being achieved" or to "feelings of desired end-state being unachieved." The former may trigger avoidance strategies while the latter may trigger adjustment behaviors.

In both cases, dysfunctional outcome processing seems to be implicated in the inconsistent experience of action. For example, neurobiological studies indirectly suggest that incompleteness phenomena are related to an impaired ability to perceive generated outcomes as being consistent with their internal representations; whereas harm avoidance and responsibility may be related to an increased focus on potential negative outcomes of action. Nevertheless, outcome processing plays an important role in the experience of action, especially the mechanism that compares events appearing upstream from action (i.e., outcome anticipation) with those occurring downstream (i.e., generated outcomes) [12-17]. Hence, theories concerning sense of agency offer a reliable context to understand the dysfunctional construal of self-action in OCD.

9.3
Sense of Agency in OCD: Empirical Data

Several models of OCD highlight a possible disturbance affecting the experience of control over one's action or events. For example, Shapiro [41] suggested that compulsions in OCD reflect a diminished inner feeling of control (sense of autonomy), leading patients to monitor their actions with a conscious effort. OCD individuals consistently self-report experiencing a diminished sense of control in everyday life (i.e., self-perception of one's ability to attain or avoid specific outcomes through one's actions) in several studies [42–44]. However, other studies showed that they can experience a high sense of control [45] or have a higher need for control [46]. These conflicting results have been interpreted as reflecting the fact that, in response to their undermined sense of control, some OCD individuals may deploy compulsions to regain control over their actions or over unwanted events [47, 48]. From this perspective, compulsive behaviors can be viewed as a way of artificially inflating affected individuals' feelings of control, when the mechanism underlying a "naturally occurring" sense of control breaks down [49, 50].

The experience of control is an important component of the sense of agency, which depends on how an individual links an action to external outcomes [14]. An action control mechanism may be crucial for assessing the extent to which one's actions produce the desired or expected outcomes; the process that compares the representation of expected or desired outcomes to observed outcomes may be the most important one in this context [12-17]. In everyday actions, people do not consciously compare "what they intended to do" with "what they actually did"; they only need to access the results of the unconscious comparison, that is, a matching signal [12, 14]. However, mismatch signals may occur when the automatic action control fails to guide and monitor an action until goal attainment (which may occur in everyday behavioral situations in most people from time to time). This, then, causes the automatic control to be passed back to a conscious monitoring of action in order to secure goal attainment [14, 49]. However, if the conscious control also fails to guide actions until the desired end-state, then it is more appropriate to abandon conscious monitoring and even to momentarily set aside the pursuit of this goal. A recent theoretical suggestion posits that, by contrast, OCD individuals are characterized by an inability to relax inefficient conscious action monitoring, leading to the deployment of a range of ways to achieve their goals [49]. To resume, the characteristic feature of OCD is an abnormally low sense of control that may be compensated through compulsive behaviors, which may have the effect of creating (artificially) a subjective experience of control. From this perspective, OCD can be reasonably viewed as a disturbance of the experience of the control component of the experience of action. The studies presented in this paragraph provide a deeper understanding about the potential defective sense of agency in OCD, through impaired action control mechanisms.

The way in which individuals with OCD symptoms understand the relationship between their actions and their related outcomes has been examined in the context of action identification theory [51, 52]. This theory posits that any behavior can be identified within a cognitive hierarchy of meanings. Higher-level meanings relate to the desired goal and expected outcomes; lower-level meanings, however, represent instrumental features and movement parameters. Vallacher and Wegner [51, 52] suggested that the particular level at which an action is identified reflects the particular representation (movement parameters *vs* outcome) on the basis of which the action is conducted and monitored. Dar and Katz [53] explored the level at which patients with washing symptoms identified the habitual act of washing their hands, compared to non-OCD controls. In their study, the authors used an item related to washing symptoms (i.e., habitually washing hands). This act was associated with 22 items varying in their level of abstraction (11 low-level items such as "I run water over my hands"; 11 high-level items such as "I show responsibility to myself"). Patients and non-OCD controls had to indicate their degree of agreement with each item. Their results suggested that patients conduct their rituals with a representation of goals and outcomes that are too abstract (e.g., "I clean myself internally," "I show responsibility to myself") compared to non-OCD controls. Furthermore, this study highlighted the unusual purpose and outcomes that are related to such a basic action. The authors suggested that OCD patients' unusual representation of the act of hand washing is related to their attempts to control potential harm. Clearly, there is no specific action plan that allows one to avoid general threatening events and disasters (e.g., preventing a fire that may destroy the building). In OCD, this fact is compensated by associating a basic action with idiopathic cues for safety; compulsions are then deployed until those cues are encountered (e.g., a specific internal state such as diminished anxiety); the identification of those cues as outcomes of one's action may then provide a feeling (albeit illusory) that one's actions can control meaningful life events.

Although the way an action is identified depends on several action-related features (e.g., action complexity, degree of expertise, action disruption or error), people tend to adopt a predominant level of action identification across behaviors (i.e., level of agency [52]). Thus, the level of agency refers to the preferential level at which actions are generally identified. People with a low level of agency tend to focus on movement parameters, including sensorimotor consequences of actions, and people with a high level of agency focus on abstract goals and on the implications of behaviors. Belayachi and Van der Linden [54] examined the relationship between OCD symptoms (i.e., checking and washing symptoms) and the level of agency (i.e., the preferential level of action identification) in non-clinical participants. In this study, participants were presented with 23 items, each of which consisted of an everyday action (e.g., locking a door) followed by two alternatives or "identities." One was a low-level depiction of the action (e.g., putting a key in the lock) and the other depicted the action at a high level (e.g., securing the house). Participants had to choose the alternative that best described each action. The results suggested that checking symptoms were related to a general tendency to identify actions mainly in terms of their procedural aspects and motor components, rather than according to the related outcomes. This is in agreement with the idea that doubts about the performance and repetition of action (which characterize checking) are related to the focus of "attention to low-level gestural units of behavior rather than to goal-related higher-level units that are normally used in action flow parsing" ([55], p. 1). Furthermore, Vallacher and Wegner [52] showed that focusing on movements during an action, and not on the goal, might impair action regulation by promoting abnormal "signals of inconsistency and error," particularly during routine actions. Overall, this fits with the idea that OCD symptoms related to incompleteness phenomena (i.e., checking) are connected with an impaired action monitoring mechanism that may inconsistently generate mismatch signals or that is not able to generate a matching signal [56]. In addition, the low level of agency was specifically related to checking symptoms, as compared to washing symptoms, which were not related to any particular level of agency. Thus, the high level of action identification observed in the Dar and Katz study seems to be specific to patients' related concerns and compulsions.

Two studies directly examined the sense of control component of sense of agency in OCD symptoms. First, Reuven-Magril, Dar, and Lieberman [50] investigated the potential relationship between the illusory experience of control, compulsive-like behavior, and OCD symptoms in both non-clinical participants and OCD patients. They used a preprogrammed sequence of neutral and aversive images for this purpose. In this task, the participants had to attempt to control and shorten the duration of the image presentation (i.e., desired outcome) by finding the right combination of key presses (i.e., action). Their perceived level of control was assessed at various points during the task. Participants did not have any actual control over the duration of the stimuli. Indeed, the presentation time gradually increased (i.e., increased discrepancy between desired and actual outcomes) throughout the first half of the task and then gradually decreased for the remaining stimuli (i.e., increased matching between desired and actual outcomes). The results showed that OCD symptoms in both clinical and non-clinical populations (regardless of OCD subtype) were related to more compulsive-like repetitive patterns of action (i.e., using the same key presses) and to an increased (illusory) sense of control for aversive and, to a lesser extent, neutral stimuli. The authors reasonably interpreted their results as consistent with the idea that compulsions must be viewed as conscious attempts to inflate OCD individuals' feelings of control over aversive events (i.e., an effortful control). Moreover, the relationship found between inflated sense of control for neutral stimuli and OCD symptoms has been related to the high need for control that characterizes some OCD individuals (i.e., the general motivation of being able to exert control over events). Furthermore, a higher illusory sense of control in OCD individuals suggests that they overestimate the extent to which their actions can control events. This is consistent with the phenomenology observed in cases of exaggerated threat anticipation and inflated responsibility, expressed as the "belief that one has a pivotal power to bring about or prevent subjectively crucial negative outcomes" ([23], p. 111).

In a later study, Belayachi and Van der Linden [57] specifically examined the mechanism that may be involved in the "naturally occurring" feeling of control over outcomes that match expectations, in non-clinical participants with OCD symptoms. The study was based on the assumption that an unconscious comparing mechanism that grasps a correspondence between expected and actual outcomes underlies the subjective experience of control (in effortless situations). This mechanism was explored with a task in which participants were made to feel that they cause an (actually uncontrollable) outcome, because this outcome was subliminally primed (emulating outcome anticipation) before the participants acted [12]. Subliminal priming of outcomes is thought to mimic the automatic activation of the representations of action effects before the action, while simultaneously preventing conscious awareness of these thoughts. In this task (the Wheel of Fortune task), the participant and the computer each moved a square in opposite directions. The participants' task was to press a key (i.e., move) to stop the motion of the squares (i.e., "actual outcome"), and subsequently to determine whether they or the computer caused the square to stop in the observed position (i.e., agency judgment). In reality, the participants had no control over the movements of the square. The outcome was arranged to not represent either the participants' or the computer's real stop position. In half of the trials, the square position to be presented was primed just before the participants stopped the motion of the square (i.e., prior thoughts about expected effect). The results showed that effect priming significantly enhanced feelings of control over the rapidly moving square, as participants experienced stronger feelings that they caused the square to stop in the presented position, on primed trials. Under these conditions, participants with checking symptoms experienced lower feelings of control in both primed and non-primed trials. This effect was found to be specific to checking, and not for the other OCD dimensions, and remained when comorbid depression was controlled for. The authors interpreted those results as being consistent with the action monitoring dysfunction hypothesis, according to which an inability to generate a consistent matching signal may lead to incompleteness phenomena. In addition, this unconscious perception of a match between expected and actual outcomes may constitute an important cue for goal satisfaction and subsequent action closure in everyday behaviors [12, 58–60]. Yet, an attenuation of this phenomenal cue may explain why checking individuals frequently experience incompleteness and doubts about the goal achievement. Thus, checking compulsions could be behavioral adjustments expressed in order to receive more convincing (explicit) cues about actual goal completion or to experience completeness (i.e., "just right" feelings). Finally, there was no association between OCD symptoms and illusory sense of control, in contrast to the link observed between illusory sense of control and OCD symptoms (regardless of OCD subtype) in the Reuven-Magril et al. study [50]. These patterns of results have been interpreted as confirming the idea that illusion of control in OCD is connected to a conscious effort to obtain such subjective effects (i.e., compulsive-like behaviors), a situation that was not allowed by the paradigm used in our study.

9.4
Summary and Discussion

Two studies have explored the way individuals with the most representative OCD symptoms (checking and washing symptoms) construe the outcomes of their actions, by assessing the level of action identification [53, 54]. The results of both studies suggest that checking symptoms are related to a tendency to identify various common actions at a low level of action construal (i.e., according to movement parameters rather than goal and outcome aspects [54]). On the other hand, washing individuals were found to identify their highly familiar act of washing hands at a higher level than non-OCD controls (i.e., a higher level of action identification [53]), although such an outcome-related identification actually reflected the fact that, compared to non-OCD controls, patients endorsed more magical and unusual outcomes depicting washers' concerns (e.g., "I clean myself internally"). This could explain why washing symptoms were not found to be related to any particularly high level of identification for various habitual actions [54]. Although comparably defective action processing has been observed in both sub-clinical and clinical OCD [11, 33, 61, 62], one could argue that the use of non-clinical participants in the Belayachi and Van der Linden study and of clinical participants in the Dar and Katz study also account for the divergent results concerning washing symptoms.

Although these data are preliminary and must be interpreted with caution, the patterns of results are rather consistent with the idea that defective outcome processing may be related to OCD, and that it may be differentially affected across OCD subtypes. Indeed, washing symptoms may be specifically motivated by harm avoidance [26], while checking symptoms may be particularly related to incompleteness experiences [26, 27]. By extrapolation, the results of the Belayachi and Van der Linden and Dar and Katz studies could suggest that incompleteness implies a lack of processing of actual outcomes generated by actions; harm avoidance, on the other hand, may be related to inconsistent processing of unrelated events that are misinterpreted as resulting from one's actions. Interestingly, the level at which action is identified may determine the extent to which people experience a feeling of control for outcomes that match expectations [63].

Consistently, the results of two studies [50, 57] that investigated the sense of control in persons with OCD symptoms can be interpreted along the same lines. Participants with OCD symptoms (regardless of OCD subtype) may be characterized by an increased illusory sense of control in compulsive-like situations (i.e., when their actions are directed towards effortful attempts to control an event [50]). However, only individuals with checking proneness appeared to experience an undermined sense of control in routine-like situations (i.e., when their actions are supposed to be automatically controlled, rather than under conscious monitoring [57]). This latter result is consistent with the van der Weiden et al. [63] study in which people with a low level of agency were also found to be less prone to experience the illusion in the Wheel of Fortune task. The authors suggested that these results reflect situations in which people have an intention to generate a specific outcome but lose track of this intention in the course of action because they did not keep a high-level representation as they monitored the action at a lower level; consequently, they may lack experience of control and agency for the intended outcomes they have yet produced. Overall, a chronic low-level of agency [54] and related undermined sense of agency [57] in individuals with checking symptoms could explain the repeated enactment of routine actions regardless of the obvious achievement of the goal.

In the study of van der Weiden et al. [63], people with a high level of action identification were found to be more prone to experience the illusion of control. This is consistent with the illusion of control in OCD individuals observed in the Reuven-Magril et al. study, but not with the absence of association between OC symptoms and an increased experience of control in the Belayachi and Van der Linden study (i.e., illusion of control, as measured by using the Wheel of Fortune task). It is rather difficult to compare these studies as the task used in each one differed in terms of the kind of experience of control it assessed. Indeed, a recent article on the contribution of sense of control to sense of agency proposed to distinguish between the "sense that one has to exert control to generate and maintain an appropriate action program" and the "sense that one feels in control of an action" ([14, p. 20]). Accordingly, the basic "sense that one feels in control of an action" only requires access to the result of the unconscious comparisons between predicted and actual states ("effortless control as for routine or automatic actions" [14]). On the other hand, the importance of mental and behavioral attempts and the adjustments necessary to reduce the discrepancies between expectations and outcomes may underlie a distinct form of experience of control (i.e., "sense that one has to exert control to generate and maintain an appropriate action program" as would be the case in disrupted or unfamiliar actions). In such effortful situations, the conscious effort itself may enhance the impression that one is engaged in and causing actions.

In light of this theoretical framework, we could argue that the illusion of control task used in the Belayachi and Van der Linden study elicited a sense of effortless control (i.e., predominantly based on automatic and unconscious processes of comparison), similar to that observed during non-conscious goal pursuit [12, 13]. Interestingly, no OCD symptoms were found to be correlated with an increased experience of control; moreover, only the checking symptoms were related to an undermined sense of control. This result is in agreement with a recent study in which self-reported low sense of control was found to be particularly related to checking symptoms [48]. In the task used by Reuven-Magril et al. [50], feelings of control were predominantly elicited by the effortful control situation created by the design of the task. Overall, this is consistent with a recent theoretical suggestion that compulsions are conscious attempts to regain control over action when automatic action monitoring fails to guide actions to the desired outcome (e.g., when the goal is too abstract [49]).

Thus, some empirical evidence seems to confirm Salkovskis's [23] assumption that harm avoidance and inflated responsibility might be construed as an illusory "perception of self-agency" It should be noted that the illusory sense of control reported by Reuven-Magril et al. [50] may highlight the phenomenon whereby compulsive-like behaviors allow OCD individuals to regain control over unwanted outcomes (e.g., aversive image) rather than the phenomenon in which they need to control, and they experience responsibility for anticipated outcomes. Indeed, a key feature of harm avoidance and responsibility phenomena is the exaggerated anticipation of negative outcomes and the belief that doing nothing to avoid those outcomes is similar to causing premeditated harm.

Interestingly, perceived premeditation may stem from any anticipation-like mental content (such as foresight, effortful forethought, wishful thinking, and the consideration of multiple possible outcomes of action). Therefore, such actions would lead people to feel responsible for those outcomes and to think that they are under personal control [64]. This counterfeit perception of self-responsibility and personal control may occur despite the obvious irrelevance of premeditation and overt behaviors, and despite the absence of any causal relationship between premeditation and observed outcomes [64]. For example, non-clinical participants can experience control and agency for observed outcomes that match prior conscious thoughts, even when the causation appears to be magical and when the thought-about outcome is viewed as undesirable [65]. Thus, future studies should explore the relationship between harm and responsibility phenomena, and the overestimation of personal influence in the occurrence of magical and/or threatening outcomes in a laboratory context.

As for incompleteness, the possible involvement of an impaired unconscious comparison mechanism fits with the idea that "not just right" experiences and incompleteness form a fragmented subjective experience resulting from the inability of action monitoring to generate consistent matching signals [56]. An alternative explanation of Belayachi and Van der Linden's [57] results is that checking, which is related to incompleteness phenomena, is linked to an abnormal access to motor and proprioceptive signals, due to low-level action identification (i.e., a predominant focus on sensorimotor features). Those signals are normally filtered and not consciously perceived [66-68]. Indeed, those signals play a minor role in our experience of action when we naturally focus on abstract outcome information allowing for goal achievement. However, a predominant focus on movements (i.e., low level of agency) may lead the experience of action to be mainly based on proprioceptive information [14, 69].

Conversely, it is possible that OCD individuals who feel a sense of incompleteness have an undamaged comparator mechanism, but their ability to interpret matching signals at a cognitive level is impaired [e.g., 59]. This may prevent them from feeling success when the expected outcome is achieved. It has consistently been demonstrated that priming knowledge of success alone can increase feelings of control to the same extent as priming of action effects [70]. Hence, it would be interesting to see whether priming the concept of success increases feelings of completeness for checking individuals. Future studies should explicitly explore this possibility in OCD individuals who experience incompleteness phenomena.

9.5
Conclusions

The symptoms of OCD can be classified according to the extent to which they are dominated by an increased sense of responsibility for random negative events (i.e., harm avoidance) or by peculiar sensory phenomena, preventing one from experiencing a sense of task completion (i.e., incompleteness). Accordingly, OCD individuals may re-enact certain actions until they obtain an outcome that informs them that they have achieved control over potential harm or until they integrate the generated outcome as consistent with the intended one. Overall, the data are consistent with the assumption that the concerns of OCD individuals regarding harm avoidance and inflated responsibility are related to an illusory "perception of self-agency" [23]. Those with incompleteness features may be characterized by an undermined sense of control, in connection with a comparator dysfunction [11, 56].

Throughout this chapter, we have tried to demonstrate how OCD can be reasonably understood as a disturbance of agency and to point out that the distinct patterns of the phenomenology of action and underlying mechanisms must be carefully studied. In both the incompleteness and harm avoidance phenomena, it is the way in which expected outcomes are compared to observed outcomes that may underlie the impaired experience of control. However, impaired unconscious comparisons could be related to a low feeling of control in incompleteness; in harm avoidance, on the other hand, dysfunctional beliefs and cognitive-related dysfunctions may account for an undermined sense of control, which also entails conscious attempts to regain control.

References

1. American Psychiatric Association (1994) Diagnostic and statistical manual of mental disorders (4th edn). American Psychiatric Association, Washington DC
2. Aouizerate B, Guehl D, Cuny E et al (2004) Pathophysiology of obsessive-compulsive disorder. A necessary link between phenomenology, neuropsychology, imagery and physiology. Prog Neurobiol 72:195-221
3. Schwartz JM (1998) Neuroanatomical aspects of cognitive-behavioural therapy response in obsessive-compulsive disorder. An evolving perspective on brain and behaviour. Br J Psychiatry Suppl 38-44
4. Schwartz JM (1999) A role of volition and attention in the generation of new brain circuitry. Toward a neurobiology of mental force. J Consciousness Stud 6:115-142
5. Rachman S (1997) A cognitive theory of obsessions. Behav Res Ther 35:793-802
6. Salkovskis P M (1985) Obsessional-compulsive problems: a cognitive-behavioural analysis. Behav Res Ther 23:571-583
7. Summerfeldt LJ (2004) Understanding and treating incompleteness in obsessive-compulsive disorder. J Clin Psychol 60:1155-1168
8. Cavedini P, Riboldi G, D'Annucci A et al (2002) Decision-making heterogeneity in obsessive-compulsive disorder: ventromedial prefrontal cortex function predicts different treatment outcomes. Neuropsychologia 40:205-211
9. Veale DM, Sahakian BJ, Owen AM, Marks IM (1996) Specific cognitive deficits in tests sensitive to frontal lobe dysfunction in obsessive-compulsive disorder. Psychol Med 26:1261-1269
10. Pitman RK (1987) A cybernetic model of obsessive-compulsive pathology. Compr Psychiatry 28:334-343
11. Gehring WJ, Himle J, Nisenson LG (2000) Action monitoring dysfunction in obsessive compulsive disorder. Psychol Sci 11:1-6
12. Aarts H, Custers R, Wegner DM (2005) On the inference of personal authorship: enhancing experienced agency by priming effect information. Conscious Cogn 14:439-458
13. Aarts H, Wegner DM, Dijksterhuis A (2006) On the feeling of doing: dysphoria and the implicit modulation of authorship ascription. Behav Res Ther 44:1621-1627
14. Pacherie E (2008) The phenomenology of action: a conceptual framework. Cognition 107:179-217
15. Wegner DM (2002) The illusion of conscious will. MIT Press, Cambridge
16. Wegner DM, Sparrow B (2004) Authorship processing. In: Gazzaniga M (ed) The new cognitive neurosciences (3rd ed). MIT Press, Cambridge, pp 1201-1209
17. Wegner DM, Wheatley TP (1999) Apparent mental causation: sources of the experience of will. Am Psychol 54:480-492
18. Jeannerod M (2009) Le cerveau volontaire. Éditions Odile Jacob, Paris
19. Rasmussen SA, Eisen JL (2002) The course and clinical features of obsessive compulsive disorder. In: Davis KL, Charney D, Coyle JT, Nemeroff C (eds) Psychopharmacology: a fifth generation of progress. Williams and Williams, Philadelphia
20. Foa EB, Kozak MJ (1995) DSM-IV field trial: obsessive-compulsive disorder. Am J Psychiatry 152:90-96

21. Rasmussen S, Eisen J (1994) The epidemiology and clinical features of obsessive-compulsive disorder. Psychiatr Clin North Am 15:743-758
22. Skoog G, Skoog I (1999) A 40 year follow-up of patients with obsessive-compulsive disorder. Arch Gen Psychiatry 56:121-127
23. Salkovskis PM (1996) Cognitive-behavioral approaches to the understanding of obsessional problems. In: Rapee RM (ed) Current controversies in the anxiety disorders. Guilford, New York, pp 103-133
24. Summerfeldt LJ, Huta V, Swinson RP (1998) Personality and obsessive-compulsive disorder. In: Swinson RP, Antony MM, Rachman S, Richter MA (eds) Obsessive-compulsive disorder: theory, research, and treatment. Guilford, New York, pp 79-119
25. Prado HS, Rosário MC, Lee J et al (2008) Sensory phenomena in obsessive-compulsive disorder and tic disorders: a review of the literature. CNS Spectr 5:425-432
26. Tolin DF, Brady RE, Hannan S (2008) Obsessional beliefs and symptoms of obsessive compulsive disorder in a clinical sample. J Psychopathol Behav Assess 30:31-42
27. Coles ME, Frost RO, Heimberg RG, Rhéaume J (2003) "Not just right experiences": perfectionism, obsessive-compulsive features and general psychopathology. Behav Res Ther 41:681-700
28. Ecker W, Gönner S (2008) Incompleteness and harm avoidance in OCD symptom dimensions. Behav Res Ther 46:895-904
29. Mataix-Cols D, Cullen S, Lange K et al (2003) Neural correlates of anxiety associated with obsessive-compulsive symptom dimensions in normal volunteers. Biol Psychiatry 53:482-93
30. Mataix-Cols D, Wooderson S, Lawrence N et al (2004) Distinct neural correlates of washing, checking, and hoarding symptom dimensions in obsessive-compulsive disorder. Arch Gen Psychiatry 61:564-76
31. Phillips ML, Marks IM, Senior C et al (2000) A differential neural response in obsessive-compulsive disorder patients with washing compared with checking symptoms to disgust. Psychol Med 30:1037-50
32. Fitzgerald KD, Welsh RC, Gehring WJ et al (2005) Error-related hyperactivity of the anterior cingulate cortex in obsessive compulsive disorder. Biol Psychiatry 57:287-294
33. Hajcak G, Simons RF (2002) Error-related brain activity in obsessive-compulsive undergraduates. Psychiatry Res 110:63-72
34. Ursu S, Stenger VA, Shear MK et al (2003) Overactive action monitoring in obsessive-compulsive disorder. Psychol Sci 14:347-353
35. Falkenstein M, Hoormann J, Christ S, Hohnsbein J (2000) ERP components on reaction errors and their functional significance: a tutorial. Biol Psychol 51:87-107
36. Gehring WJ, Goss B, Coles MGH et al (1993) A neural system for error detection and compensation. Psychol Sci 4:385-390
37. Kringelbach ML (2005) The human orbitofrontal cortex: linking reward to hedonic experience. Nat Rev Neurosci 6:691-702
38. O'Doherty J, Kringelbach ML, Rolls ET et al (2001) Abstract reward and punishment representations in the human orbitofrontal cortex. Nat Neurosci 4:95-102
39. Ursu S, Carter CS (2005) Outcome representations, counterfactual comparisons and the human orbitofrontal cortex: implications for neuroimaging studies of decision-making. Brain Res Cogn Brain Res 23:51-60
40. Ursu S, Carter CS (2009) An initial investigation of the orbitofrontal cortex hyperactivity in obsessive-compulsive disorder: exaggerated representations of anticipated aversive events? Neuropsychologia 47:2145-2148
41. Shapiro D (1965) Neurotic styles. Basic Books, New York
42. Freeston MH, Ladouceur, R (1997) What do patients do with their obsessive thoughts? Behav Res Ther 35:335-348

43. Ladouceur R, Freeston MH, Fournier S et al (2000) Strategies used with intrusive thoughts: a comparison of OCD patients with anxious and community controls. J Abnorm Psychol 109:179-187
44. McLaren S, Crowe SF (2003) The contribution of perceived control of stressful life events and thought suppression to the symptoms of obsessive-compulsive disorder in both non-clinical and clinical samples. J Anxiety Disord 17:389-403
45. Rhéaume J, Ladouceur R, Freeston MH, Letarte H (1995) Inflated responsibility in OCD: validation of an operational definition. Behav Res Ther 33:159-169
46. Sookman D, Pinard G, Beck AT (2001) Vulnerability schemas in obsessive-compulsive disorder. J Cogn Psychother 15:109-130
47. Moulding R, Kyrios M (2006) Anxiety disorders and control related beliefs: the exemplar of obsessive-compulsive disorder (OCD). Clin Psychol Rev 26:573-583
48. Moulding R, Kyrios M (2007) Desire for control, sense of control and obsessive compulsive symptoms. Cogn Ther Res 31:759-772
49. Liberman N, Dar R (2009) Normal and pathological consequences of encountering difficulties in monitoring progress toward goals. In: Moskowitz G, Grant H (eds) The psychology of goals. Guilford, New York, pp 277-303
50. Reuven-Magril O, Dar R, Liberman N (2008) Illusion of control and behavioral control attempts in obsessive-compulsive disorder. J Abnorm Psychol 117:334-341
51. Vallacher RR, Wegner DM (1985) A theory of action identification. Lawrence Erlbaum Associates, Hillsdale, NJ
52. Vallacher RR, Wegner DM (1989) Levels of personal agency: individual variation in action identification. J Pers Soc Psychol 57:660-671
53. Dar R, Katz H (2005) Action identification in obsessive-compulsive washers. Cogn Ther Res 29:333-341
54. Belayachi S, Van der Linden M (2009) Level of agency in sub-clinical checking. Conscious Cogn 18:293-299
55. Boyer P, Liénard P (2006) Why ritualized behavior? Precaution systems and action parsing in developmental, pathological and cultural rituals. Behavioral and Brain Sciences 29:1-56
56. Coles ME, Heimberg RG, Frost RO, Steketee G (2005) Not just right experiences and obsessive-compulsive features: experimental, self-monitoring perspectives. Behav Res Ther 43:153-167
57. Belayachi S, Van der Linden M (in press) Feeling of doing in obsessive-compulsive checking. Conscious Cogn
58. Szechtman H, Woody EZ (2004) Obsessive-compulsive disorder as a disturbance of security motivation. Psychol Rev 111:111-127
59. Woody EZ, Lewis V, Snider L et al (2005) Induction of compulsive-like washing by blocking the feeling of knowing: an experimental test of the security-motivation hypothesis of obsessive-compulsive disorder. Behav Brain Funct 1:1-11
60. Woody E, Szechtman H (2000) Hypnotic hallucinations and yedasentience. Contemp Hypn 17:26-31
61. Ecker W, Engelkamp J (1995) Memory for actions in obsessive-compulsive disorder. Behav Cogn Psychother 23:349-371
62. Zermatten A, Van der Linden M, Larøi F, Ceschi G (2006) Reality monitoring and motor memory in checking-prone individuals. J Anxiety Disord 20:580-596
63. van der Weiden A, Aarts H, Ruys K (2010) Reflecting on the action or its outcome: behavior representation level modulates high level outcome priming effects on self-agency experiences. Conscious Cogn doi:10.1016/j.concog.2009.12.004
64. Morewedge CK, Gray K, Wegner DM (in press) Perish the forethought: premeditation engenders misperceptions of personal control. In: Hassin R, Ochsner K, Trope Y (eds) Self-control in society, mind, and brain. Oxford University Press, New York

65. Pronin E, Wegner DM, McCarthy K, Rodriguez S (2006) Everyday magical powers: the role of apparent mental causation in the overestimation of personal influence. J Pers Soc Psychol 91:218-231
66. Blakemore S-J, Wolpert DM, Frith CD (2002) Abnormalities in the awareness of action. Trends Cogn Sci 6:237 -242
67. Fourneret P, Jeannerod M (1998) Limited conscious monitoring of motor performance in normal subjects. Neuropsychologia 36:1133-1140
68. Frith CD (2005) The self in action: lessons from delusions of control. Conscious Cogn 4:752-770
69. Wohlschläger A, Engbert K, Haggard P et al (2003) Intentionality as a constituting condition for the own self and selves. Conscious Cogn 4:708-716
70. Aarts H (2007) Unconscious authorship ascription: the effects of success and effect-specific information priming on experienced authorship. J Exp Soc Psychol 43:119-126

Body and Self-awareness: Functional and Dysfunctional Mechanisms 10

M. Balconi, A. Bortolotti

10.1
The Sense of Agency and the Sense of Ownership as Components of Self-consciousness

Some features of human experience contribute to a person's self-consciousness as the "ability to represent one's own bodily and mental states as one's own states" [1]. Although some aspects of this ability are phenomenologically the same, they are heterogeneous on both the functional and the representational level. Experienced phenomena involved in self-consciousness are the sum of one's own experiences, the perspectivity of these experiences, the sense of ownership of one's own body parts, the sense of agency of actions, the sense of authorship of thoughts, and the trans-temporal integration of all this into autobiographical knowledge [1]. These aspects highlight the psychological, physiological, and neural mechanisms involved in bodily experience and important for self-consciousness.

The agency of actions can be defined as one form of bodily self-consciousness, because it contains the idea that someone perceives him/herself as the agent of action [2]. By extension, the sense of ownership of one's body parts (body ownership) is defined as the feeling that my body belongs to me [3]. Body ownership therefore contributes to self-consciousness since it represents the continuous experience related to the feeling of belonging to one's own body which, in turn, contributes to self-representation.

According to Gallagher [3], the self can be defined through two different concepts representing two aspects of self: the minimal self and the narrative self. The *minimal self* is self-consciousness as an immediate subject of experience without temporal continuity; it can be seen as a pre-reflective point of origin for action, expe-

M. Balconi (✉)
Department of Psychology, Catholic University of Milan, Milan, Italy

rience, and thought as well as a privileged veridical knowledge about what is identifiable with the pronoun "I." This aspect of self means that we cannot make a mistake about the person to whom we are referring to (*immunity principle*) [4]. It is limited to what is accessible to immediate self-consciousness since it is not conceptual and is linked to a first-person perspective. Cognitive processes do not mediate the first-person experience with known criteria for judging the experience of being myself. However, not all thoughts about the self can be identified for their independence with respect to semantic identifications or identity criteria. Not only can the same property be attributed to oneself through different ways of knowing [5], but there is also a difference between having a mental or physical state and the acknowledgment that one has that state.

The second, *narrative self* is the self-image that originates from the stories of oneself; it is linked with the past and the future and is therefore extended in time [3]. This form of self allows past and future actions to confer a general coherence in terms of goals, causes, projects, and general guidelines. The narrative self was defined by Dennett as a "center of narrative gravity," with the constant purpose to build models, as representations of self, and to make variously modifiable connections. The concept of narrative self emerges in psychology from Neisser's concepts [6] of *extended self* and *conceptual self*, the former connected to significant past experiences and future expectations and the latter related to the set of assumptions or sub-theories concerning social roles, body, mind, and self-assigned features. These concepts, originally explained in terms of memory, have earned renewed consideration in the context of language and the role of narrative in the development of one's own self-concept [6]. Moreover, neuropsychological evidence that mental processes are distributed across various brain regions does not support the possibility of a real center of experience. However, broadly speaking, the left hemisphere is responsible for generating narratives, using the *interpreter* [7]. The autobiographical fact and the inventive fiction are together considered by this mechanism that produces a sense of a continuous self. The interpreter does not produce a false self because it makes sense of what actually happens to the person. Gallagher [3] proposed a model in which the extended self is decentered and distributed and conflicts such as moral indecision and self-deception are considered (see also Chapter 1).

10.2
The Sense of Body Ownership *vs* the Sense of Agency

The minimal level of self-awareness is related to two other aspects of self: the *sense of agency*, which is the sense that a particular someone is causing an action, and the *sense of ownership*, or the sense that a particular someone is undergoing an experience. The sense of self-ownership is not vulnerable to the influence of predicted and actual feedback of actions, unlike the sense of self-agency. This fact supports the idea of the independence of these two senses of self [8].

Synofzik and colleagues [1] defined the sense of agency and the sense of ownership as phenomenal experiences of the "mineness" of actions that are present in the non-conceptual representation of oneself as agent. The sense of agency and the sense of ownership coincide and are indistinguishable in the normal experience of voluntary action. Nevertheless, in passive movements, the sense of agency and the sense of ownership are dissociated: the sense of ownership is implicated in the sense of agency but not vice versa, because only the sense of agency needs self-generated movements [9].

Thus, body ownership is present not only during voluntary actions (sense of agency) but also during passive experience. This implies that the sense of ownership triggers different forms of body awareness, depending on the presence or absence of the sense of agency [9]. In particular, the sense of body ownership is the experience that allows us to affirm "that is my movement" or "I'm the subject of the movement" [10]. It is a pre-reflective sense that does not require an explicit, observational consciousness of the body or an intentional consideration of the body as an object. The body is not regarded as an object of experience but it does represent the subject of experience [11]. This concept is evident in the finding that judgment of agency does not include the process of recognition of oneself as oneself [10]. Bodily self-knowledge, in the sense of ownership, is identification-free, relies directly on the proprioceptive system and therefore is immune to error through mis-identification [12]. In fact, proprioceptive self-ascription does not demand identification of the body as one's own, unlike the visual self-ascription of bodily properties. The sense of ownership can be defined as the first-personal perspective that allows me to feel that my own body is the source of sensations [13]. According to Gallagher [3], body ownership is the feeling of one's body belongs to oneself, and this feeling is present in human mental life.

The sense of agency and the sense of ownership can be distinguished both in first-order experience and in higher-order consciousness [9]. Nevertheless, the sense of ownership, as the sense of agency, has a representational structure, a multi-level framework that gradually increases in representational and functional complexity: from basic non-conceptual sensorimotor representations to conceptual representations of agency and ownership [1]. Both phenomena originate from a combination of afferent sensory signals and efferent motor signals [9]. The sense of agency and the sense of ownership are probably generated by low-level sensorimotor processes that enable a person to be aware of momentary representation of one's ownership of body parts. This is possible because a system registers sensory inputs as the inputs of one's own body [1].

10.3
The Sense of My Body as Mine: A Threefold Perspective

In the individual's relationship with the world, information regarding one's own body and the position of objects in space is processed through various sensory modalities. This information is retained in the person's mind in terms of his own position and those of the considered objects. The basis of this ability is the perceptual processing of signals from different senses and the planning and execution of motor acts. Body and objects are distinguishable with respect to their processing as special patterns of intersensory information [14].

According to Tsakiris and coworkers [1], it is possible to distinguish among different forms of bodily self-representations: feeling (perceptual representation of body), judgment (propositional representation of the body), and meta-representation of the body (Fig. 10.1). Thus, at the level of *feeling of ownership*, a non-conceptual representation of one's own body can be built in response to proprioception and visual feedback, which are integrated with a *pre-existing body schema*. Low-level sensorimotor processes, understood as "ownership indicators" or bottom-up inputs, allow a person to represent his self as owner of his own body parts. This aspect of the sense of ownership also includes a non-sensory top-down component that mediates the sensory bottom-up inputs [15]. If ownership indicators are congruent, the experience of

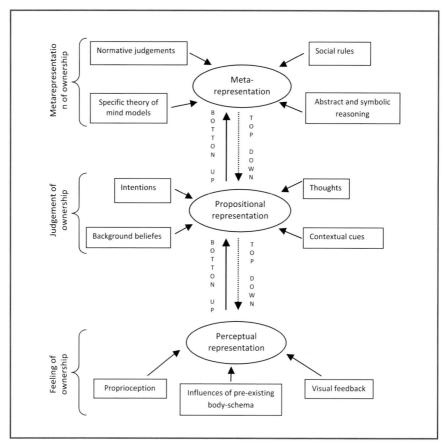

Fig. 10.1 Levels of the sense of ownership as a form of bodily self-representation. (Partially modified from Synofzik et al. [1], with permission)

bodily ownership will confer a diffuse feeling of a coherent, ongoing flow of bodily experiences. The experience of being the owner of one's body occurs beyond consciousness. It can be defined as a feeling of ownership in which neither internal prediction nor the comparator mechanism is triggered, unlike the feeling of agency [1].

The idea that we can have a sense of feeling of ownership refers to Damasio's concept of "background feeling" as a temporary state of the organism, a perception of the state of the body as a whole [16]. According to Damasio [16], conscious knowledge that each part of my body is my own is not constantly renewed, but it is impossible to think that the corporal state assumes a permanent place in one's consciousness. The sense of body is always present, even if it is not noticeable: it is not a specific part of something, but a comprehensive state. It is possible that we perceive the corporal state of ourselves every time because of a conscious re-mapping of proprioceptional information. The continuity of this feeling is linked to the idea that the human organism, its structure, and the feeling of both are continuous throughout life.

It is possible that there is an explicit *judgment of body ownership* and it defines an interpretative assessment about the awareness of being the owner of one's own body. On this conceptual level, different cognitive ownership indicators concur to yield a definition of bodily ownership as a processed component. Conceptual capacities, belief stances, intentions, and contextual cues accompany an interpretative judgment of body ownership and they permit the process of ownership attribution. The judgement of ownership is important since its disruption leads to neurological pathologies, such as alien limb phenomena or somatoparaphrenic delusional belief [1].

The third level, the *meta-representation of ownership*, illustrates the ability of the sense of body ownership to extend itself. Certain ideologies, background beliefs, and socio-cultural norms are not proper parts of body but can be seen as extensions of body ownership. In fact, when used extensively, these "objects" could be integrated into the body schema, implying that the meta-representation of ownership is an extension of the body's boundaries beyond its physical self to include socio-normative attribution. The meta-representational level allows us to introduce a discussion about the brain's potential to include certain extracorporeal objects that maintain a systematic relationship with the body itself. Plastic changes of body image are not, however, the product of passive changes that take place in response to proprioceptive, kinesthetic, or visual input per se [17]; instead, they are tightly linked to an active and purposeful use of certain "tools" as physical extensions of the body.

Although the sense of ownership is determined by the voluntariness of a physical act, visual and proprioceptive afferent signals and efferent motor signals are important for the experience of being the owner of one's own body [18]. Afferent and efferent aspects of the moving body can influence the sense of ownership [13], with efferent information playing a supplementary role in the constitution of a coherent bodily experience. This effect emerges from research conducted by Tsakiris and colleagues [13]. Their results showed a similar effect for two different experimental conditions within a kinesthetic paradigm equivalent of the rubber hand illusion (RHI), in which the perceived position of one's moving finger drifts towards a synchronous video image of the hand (seen on a 15-cm video screen) far from the real hand.

In this experimental paradigm, subjects are confronted with two different conditions: an active condition, in which the subject's finger is moved actively by the subject himself, and a passive condition, in which the same action occurs passively, with the subject's finger being moved by the experimenter. While in the passive condition only the stimulated finger is perceived to be closer to the video-projected hand, in the active condition not only the stimulated finger but also the unstimulated index finger is perceived as being closer to the projected hand image. These data suggest that efferent information of the active condition does not elicit fragmented body ownership but integrates the bodily experience. It is possible to argue that there are multisensory and sensorimotor mechanisms for the global representation of the body and that they are the outcome of integrated processes, including those extending beyond representations of certain body parts [18]. Efferent information influences body ownership since it unifies the body as whole [19].

10.4
A Spatial Hypothesis of Body Representation

The perception of one's own body is an essential requirement for one's interactions with the outside environment and it can be seen as contributing to self-consciousness [20]. In fact, the basis of spatial cognition is to be found in the constant interaction of the human body with the world in which it is inserted. As noted above, spatial cognition keeps track of the body's position and that of objects. The basis of this ability is the perceptual processing of signals from different senses together with the planning and execution of motor acts. This sensorimotor information give shape to internal representations of one's own body and of the objects around it in space.

If body ownership can be defined as the immediate and continuous experience of the feeling that one's own body belongs to oneself [21], then *embodiment* can be seen as the experience in which the self is localized within one's own body and at a certain position in space [22]. Bodily experiences are made up of *descriptive content* and *spatial content*. The former refers to the body proper as consisting of the component body parts while the latter assigns a specific location each part [23]. According to De Vignemont [19], the sense of ownership derives from the spatial representation of the body, from the localization of a bodily sensation on a map of the body. This view can be applied to define the sense of embodiment.

This idea of the sense of ownership was expressed by Martin [24] as the bodily experience of the boundaries of one's own body: the spatial structure of bodily sensations is connected to *somatosensory experience* within the boundaries of the body. However, this view is limited because not only somatosensory representation of one's own body can emerge from bodily experience [19]. The alternative is to understand bodily experiences as being based on a *multimodal representation of the body* that is used to designate a set of interconnected perceptual and motor functions essential for the performance of other functions, including the perception and localization of somatic stimuli, the planning of actions, and body awareness [25]. Against this background of a multimodal representation of the body, it may be possible to understand bodily sensations [19]. The consequences for the sense of ownership are linked to the consideration that it results from the tactile property within body representation, which is constructed on the basis of the available information (from vision, touch, or proprioception).

Nonetheless, although the sense of ownership results from bodily sensations, this does not mean that it cannot be influenced by other sources of information. The body can be an object viewed from different perspectives and described in terms of its different properties. The above-mentioned RHI is evidence of this multimodal characterization, because it demonstrates that vision is an important aspect for the bodily experience of ownership (see also Chapter 6, Par. 6.1). Since the body schema is embedded in one's body perception and the sense of ownership is given by the body's spatial content, the body schema can be seen as contributing to the sense of ownership [19]. The body schema, in fact, can be identified with an unconscious functional sensorimotor map of the body that provides the essential information needed to move one's own body. In this map, bodily contents are aspects of the relationship between the various parts of the body, individually and as a whole. Since it represents only aspect that it considers as belonging to one's own body, the body schema is first-personal.

Moreover, the body schema represents the acting body because of its involvement in performing an action and in reaching a goal. In this view, the map of the body, as part of the body schema, could be the basis of ownership because it can localize bodily sensations. This is supported by the evidence that patients with "numbsense," i.e., lacking a sense of touch for a particular body part, can localize sensation accurately, pointing to the location of the touch. The explanation for this may lie in the RHI paradigm and, relatedly, the role of the body schema for the sense of ownership. According to De Vignemont [19], the RHI occurs because of an illusory sense of ownership of the rubber hand. This conclusion implies that a representation of body movement can be used to elicit a sense of ownership, and a representation of the image of the hand included in the body schema can be used to achieve the same effect, i.e., a sense of ownership of the hand. The role of the body schema in the sense of ownership is also demonstrated by the experiences reported by some people with a prosthetic limb. The prosthesis experience can generate a sense of ownership in which the artificial limb substitutes for the missing limb in the body schema. This adoption of the prosthesis determines a feeling of ownership in a way that is useful for motor control.

Carruthers [26] criticized de Vignemont's model, stating that, while this paradigm may be regarded as model of the self-conscious sense of embodiment, it is not a model of the experience of one's body attributed to oneself nor does it show how the integration of bodily maps into the body schema is sufficient to elicit a sense of embodiment in a virtual hand or prosthetic limb. As an alternative model, Carruthers [26] proposed an offline representation of body image as underlying the sense of embodiment. Indeed, embodiment, primarily used for self-recognition, is based on both *online* and *offline representations* of the body. These two different representations of the body experience differ in terms of their origin and content. Online representations are generated by sensory input and reflect the body's current status. They allow online control and monitoring of body configurations. Offline representations are built by online representations and reflect the way in which the body is usually. Thus, they encode more permanent aspects of the body, such as its structure and its typical motor features. Carruthers argued that only offline body representations underlie the sense of embodiment. Starting from several body-related neurological and psychiatric conditions, such as anosognosia for hemiplegia and body integrity identity disorder, he defined the sense of embodiment as the result of the integration of various types of offline representations with visual and emotional information. By contrast, online representations are not necessary and are not sufficient for embodiment.

Carruthers' critique of de Vignemont's model provides further insight into spatial hypotheses of body representation, not that it is also exempt from critique. Indeed, Tsakiris and Fotopoulou [27] expressed several doubts about the Carruthers' model [28]. They highlighted that embodiment is not primarily for the purpose of self-recognition because it also plays a role in the process of distinction between self and the external world. Moreover, they argued that online/offline representations are more complex. First, the idea of a "direct" process in online representation must consider that multisensory perception is not merely a registration of peripheral inputs but involves the interpretation of sensory inputs in the context of a rich multisensory model of the body and the use of these inputs for an online representation of the body in space [29]. Second, it is possible to form an offline representation of the body without a previous online experience or sensation. Third, it is incorrect to assume that offline representations cannot modulate online representations; for example, in the case of a phantom limb, peripheral information is interpreted with reference to an offline model of the body's structure that continues to exist even in the absence of the limb. According to Tsakiris and Fotopoulou [27], the Carruthers' model is incomplete because it avoids the role of temporality and spatiality by ignoring the sense of our own body.

10.5
Neural Substrates of the Sense of Ownership

As a neural and functional field of research, the feeling of body ownership is easily investigated [9]. By contrast, the construct of body ownership is empirically difficult to isolate and it may be confused with the sense of controlling one's own body, because agency corresponds to a strong element of body ownership. The results of neuroimaging studies [30, 31] have underlined several neural correlates that account for our own experience as one source of action and the experience of knowing someone else as another. According to Farrer and Frith's experiment [30], in which an action was identified as being either one's own or that of an experimenter, to be aware that I am the cause of an action is associated with anterior insula activation; while the awareness that I am not the cause of the action, i.e., attributing the action to another person, likely involves activation of the parietal inferior cortex. This structure is thought to represent movements in an allocentric coding system that can be applied to our own actions and those of others.

The network implicated in the sense of ownership includes the somatosensory cortex, the posterior parietal lobe, and the insula. Kinesthetic and emotional information functioning related to homeostasis have been ascribed to the insula [32], while more cognitive aspects, such as spatial organization and the general semantics of the body, the relationship between the psychic self and the material body, and the distinction of one's own body from that of others, are fulfilled by the posterior parietal cortex [33, 34]. Thus, the parietal lobe may be involved in the experience of sense of agency since it is able to detect movements and to distinguish those made by ourselves from those made by others [35]. Specifically, the superior portion of the parietal lobe probably plays a role in maintaining one's own body image. It integrates synchronized visual and proprioceptive inputs to update the body image, while the inferior parietal area detects the movements of others [34]. Tsakiris and colleagues [9] showed that activation of the right posterior insular cortex during the RHI represents a sense of body ownership. Activation of the anterior insular cortex has been implicated in human awareness. This structure is thought to be associated with a somatotopic representation of feeling arising from movements as part of a wider representation of all feelings related to and stemming from one's own body [36]. It is also involved in emotional aspects of body consciousness since insular lesions can cause somatic hallucinations [37] and electronic stimulations near the insular cortex can induce illusions of changes in body position as well as feelings of being outside of one's own body [38].

Farrer and colleagues [31] detected activation of the posterior insula in processes of self-attribution and, in particular, in correlation with the degree of congruence between the different signals used to assign an action to oneself or to another. Tsakiris and colleagues [9] recorded activation also in the absence of efferent information and in correspondence with the integration of multisensory information in the process of attribution of one's body parts to one's own self. The involvement of efferent signals is unnecessary and only cortical activation is connected with body ownership. The latter emerges as a form of self-attribution for body parts, consequently, implicating the insular cortex as a key component of bodily self-awareness [9].

In an fMRI study and using the experimental paradigm of the RHI, Ehrsson and colleagues [39] showed activation of the bilateral ventral premotor cortex, left intraparietal cortex, and bilateral cerebellum. In particular, it would appear that the detection of correlated multisensory signals mediates the feeling of body ownership, due to the role of the premotor cortex and cerebellum [40]. The premotor cortex prepares the postural muscles for the start of the movement and orients the body and the arm toward a stimulus [41]. Evidence for these functions is the large afferents from the posterior parietal cortex and prefrontal lobe that provide the body with information about its spatial orientation, and the cerebellum. In particular, the ventral premotor cortex has cognitive and motor functions, with the latter determining the transition from recognition of the intrinsic properties of an object to actions of the hand, and from spatial localization to actions of the hand and arm [42]. Half of the motor neurons of the ventral premotor cortex respond, in fact, to somatosensory stimuli and approximately a fifth of them show visual feedback [43]. The more strictly cognitive aspects of this cerebral area concern the perception of space, the understanding of action, and imitation [42]. Bilateral activation of the premotor cortex seems to be linked to the feeling of ownership, particularly of the hand, based on correlations that are made with different sources of information [40]. This area of the brain, anatomically connected with somatosensory areas, receives visual, tactile, and proprioceptive inputs in addition to achieving multisensory integration of this information. Cerebellar activation occurs in the analysis of sensory information, including the integration of different kinds of information about body representations [40]. The cerebellum is not extensively involved in the detection of synchrony [39] but in processing tactile information from two matched body parts, at the level of detection of temporally correlated signals.

Tsakiris and colleagues [9] pointed out that we can adduce a further cerebral distinction, in which attention is focused on conditions that induce the RHI or on the lasting effect of sense ownership itself. Incorporation of a rubber hand into one's own body representation is reflected in activation of the right posterior insula and right frontal operculum while the feeling of ownership related to the failure of this experience activates the contralateral primary and secondary somatosensory cortices. In that study, the premotor area was not activated, in contrast to the findings of Ehrsson's group. This discrepancy may be explained by considering whether or not both the sense of body ownership of the rubber hand and the conditions that lead to it were included in the analysis. Thus, Ehrsson and colleagues considered the onset of the incorporation process, and found activation of the premotor area. The results of Tsakiris' group suggest that the focus of that study was the steady state of incorporation, in which case the activity of the premotor cortex would be unclear. These findings highlight the need for more detailed studies, during a longer period of stimulation or, conversely, in which the analysis is limited to the onset of those processes that lead to the experience of body ownership.

10.6
Disruption of the Sense of Ownership: Conscious and Non-conscious Body Perception

Body cognition is bound to the integration of several levels of nervous activity, from the analysis of primary somatosensorial afferences to the more complex processing of self-awareness [25]. Some anatomico-functional features of body representation are useful to understanding how certain neurological disorders and neuropsychological pathologies damage the correct awareness of one's own body [44].

The first feature of body representation is the *integrity of primary sensorial afferences*, mainly concerning proprioception. Therefore, localization of a body part requires not only a combination of afferent information but also a stored representation of the body [45]. The second feature is the *multisensorial nature of body repre-*

sentation: the cognitive processes linked to somato-representation contribute to a semantic knowledge and a distinct attitude about the body. Thus, somato-representation includes the *lexical-semantic knowledge* about one's own body and the body in general, *configural knowledge* about the body's structure, the *emotional and attitudinal aspects* of one's own body, and the link between the *physical body* and *psychological self* [45].

Examples of experiences that contribute to the comprehension of the importance of our own corporeal state include *autoscopic phenomena* [46], in which there is a failure to integrate multisensorial information from one's own body, resulting in a feeling of disembodiment and the impression of seeing the environment and one's own self from an elevated and distant visuo-spatial perspective [47]. The entire body can be disturbed systematically [48], including autoscopic hallucinations, heautoscopy, out-of-body experience, and feeling-of-a-presence. In *autoscopic hallucinations*, patients see a double of their own body in extrapersonal space but they do not self-attribute the illusory body or self-localize themselves at its position, as happens in patients with heautoscopy [46]. In the case of *out-of-body experiences*, the localization and attribution of the self with an illusory extracorporeal body are complete: patients see their body from a disembodied location. Finally, in the experience of *feeling-of-a-presence* there is not a visual illusion of one's own body, but the illusory body is experienced as the body of another human.

Autoscopic phenomena differ from each other in some respects. Thus, out-of-body experiences and feeling-of-a-presence but not autoscopic hallucination can be considered as vestibular disturbances. While all three phenomena are linked to a disintegration of personal space, only out-of-body experience and heautoscopy show a disintegration of both personal and extrapersonal space. Consequently, out-of-body experience and heautoscopy can be regarded as disorders in embodiment and body ownership; while the feeling-of-a-presence is strictly a body ownership disorder, in which the main brain regions involved are the premotor area, posterior parietal areas, and the temporo-parietal junction [46, 49]. These phenomena demonstrate the importance of vestibular and multi-somatosensory processing in coding for embodiment and body ownership.

Another essential feature of body representation is its *stability* and, as evidenced by the phenomenon of a phantom limb (conscious persistence of the perception of a limb or body segment, despite its mutilation, or the congenital absence), it can be altered as well. Moreover, people who suffer amputation of a limb usually report that the phantom limb can change in shape and size over time. Their experience of a phantom limb can include the ability to describe in detail its posture and to move it voluntarily, as well as to report uncontrollable and painful sensations coming from it [50].

In contrast to this stable representation of the body despite de-afferentation, some patients with brain lesions can testify to the presence of multiple body parts, in most cases hands or feet [51]. The sensations are vivid, precise, and may differ depending on the location of the cerebral lesion [25]. In cases involving parietal lesions, especially in the right hemisphere, the perception of a supernumerary limb might be related to a somatosensory deficit combined with a lack of awareness. In patients with a fronto-mesial lesion, there may be persistence in the activation of a premotor region following completion of a movement, resulting in the sensation that a supernumerary limb is carrying out a movement, although the movement was already accomplished. Claims of supernumerary limbs can coexist with a denial of hemiplegia and of the contralesional limb, suggesting that negative (e.g., of not belonging to the body) and positive (e,g., surplus limbs) symptoms arise from common mechanisms.

10.6.1
The Rubber Hand Illusion: Evidence of Disownership Phenomena

As discussed throughout this chapter, brain mechanisms involved in the coding of self-attribution for body parts in healthy subjects can be examined using the RHI. During this experimental condition, erroneous self-attribution of the fake hand is associated with errors in the localization of one's own hand [15]. This paradigm is therefore a useful model to isolate the pure sense of body ownership, in the absence of movement and efferent information [9]. It would, therefore, release the sense of body ownership from the sense of agency as these two senses coincide in voluntary actions. Whereas body ownership shows the importance of interactions between vision, touch and proprioception, the sense of agency requires the integration of information concerning the body, the world, and efferent signals. The neural processes implicated in the sense of agency are the same as those responsible for the motor aspects of action [10]. In the RHI paradigm, the passivity of the subject, i.e., the lack of a sense of agency, can be exploited to bring out the sense of body ownership and to reveal the basis of bodily self-identification.

The RHI requires the congruency of visual-tactile and object stimulation with a pre-existing representation of one's own body, as the synchronous stimulation of the real hand, hidden from view, and the rubber hand seems to produce coordination of what subjects see with what they hear [52]. This experimental condition results in displacement of the subject's hand from the position in which he or she believes it is in the direction of the artificial hand. The duration of the illusion is related to the sensation of movement. The greater the visual and tactile synchronization, the more the rubber hand is perceived as being one's own [15].

The phenomenon of the RHI can be explained as an overlap between proprioceptive and visual input [14]. This overlap seems to result from sensory feedback related to the self and it provides us with an awareness of the spatial position of specific body parts. However, the phenomenon is not observed if the experimental conditions are arranged such that there is stimulation of the left hand, with the rubber hand on the right [15]. This argues in favor of the idea that the correlation of tactile and visual perception is necessary but not sufficient for the sense of body ownership, which for the RHI would require that the rubber hand moves within a general pre-existing representation of body scheme [9]. A conflict between tactile and visual perception causes activation of the right frontal cortex, which monitors the perception of body-related sensory signals, and disappearance of the attribution of the rubber hand to one's own body [9]. The overlap only of the visual and proprioceptive maps is not sufficient to generate the RHI. Instead, the modulation of body ownership, as seen in the RHI, may require the influence of top-down visual, proprioceptive, and functional processes of body representation [15].

Moreover, active movements appear to modulate body ownership, or rather the localization of one's body's parts, beyond the stimulation applied in the RHI [13]. Body awareness is linked to proprioceptive awareness, which includes the conscious experience of the location of a specific body part in space [13] and the definition of body boundaries. Proprioceptive awareness allows us to explore objects and it guides our movements. However, the feeling derived from seeing the rubber hand and the tactile stimulation applied in the RHI involves more than just tactile and visual sensations: there is a persistent phenomenological change in the representation of the body. Thus, the RHI could generate an interaction between general body-scheme representations and localized visual and tactile integration [15].

The interaction of the body schema in the sense of body ownership can be considered in light of the results of Ehrsson and colleagues [39]: if the rubber hand is oriented 90° or 180° to the hand of the subject, the illusion fails. This result suggests that body ownership derives (at least in part) from its integration with the body schema, that is, the model of one's own body as an entity capable of assessing postures and movements. Body ownership, then, would seem to require integration of the interconnected perceptions and motor functions that are essential for the performance of various functions, including the perception and localization of somatic stimuli, the planning of actions, and body awareness [25]. Ultimately, if the object of experimental condition is a shapeless piece of wood and not a rubber hand, the illusion is reduced [15], suggesting that body ownership also requires integration of an appropriate body image.

10.6.2
Other Body Impairments: Neuropsychological Disorders

Neuropsychological disorders about body ownership include the incapacity to point to specific body parts (autotopagnosia) or, more commonly, a lack of body awareness (somatoparaphrenia, unilateral personal neglect, or hemisomatoagnosia).

Autotopagnosia is an unusual clinical disorder that is caused by focal lesions in the left parietal or occipito-parietal regions, specifically, in the language-dominant hemisphere [25]. In the experimental paradigm of Denes and colleagues [53], patients with autotopagnosia were asked to describe pictures of body parts or solid objects located in different positions and shown sequentially. Performance was poor only for the task requiring a description of body parts, suggesting that this disorder is linked to the incapacity of conscious access to the representation of spatial relations between body parts. Another study [54] reported similar findings in an experimental paradigm in which the patient was asked to compare pictures in which body parts and objects were placed at different angles, and visually degraded pictures of body parts and objects. Performance was better for the description of objects in both tasks, leading the authors to conclude that autotopagnosia arises from a lack of access to the structural description of the human body.

Schwoebel and colleagues [55] underlined, through a single case study, some interesting aspects of autotopagnosia: (1) very poor localization of tactile stimuli, (2) the proximity of the touched body part as an important factor in the sequential discrimination of stimuli, and (3) rotation of the body axis as a negative factor in the patient's ability to touch the body at the stimulated point. The authors argued that their patient with this disorder was unable to code the position of body parts in an egocentric coordinated system; instead using an esocentric coordinated system. Also, simple tasks involving the body need representational mediation, require spatial and functional interpretation, and are independent from the representational mediation for complex objects. This representational mediation is localized in the parietal lobule and lateralized to the same side as language. Thus, patients with autotopagnosia have access to the semantic meaning of the various body parts, are aware of their body, and are capable of using the body surface as a spatial map, but fail to locate the spatial position of individual body parts [25].

An additional example of neuropsychological impairment characterized by poor body ownership without an explicit link to the representation of extrapersonal space is *personal neglect*. This disorder, in the form of hemisomatoagnosia, suggests a specific alteration in the body schema [56] and can thus be viewed as a body awareness disorder. Patients with hemisomatoagnosia have a propensity to ignore the contralesional side of their own body [25]. Some typical clinical signals of this disorder are related to the tendency to act like the left side of the body is nonexistent (in lesions to the right hemisphere, which is the most common form), such that patients assume unusual body postures or fail to dress and to care for the left side of their bodies.

Patients with *somatoparaphrenia* also show body ownership disruption. The disorder is defined as an alteration of awareness regarding the involved body part (often located on the contralateral side of the injured hemisphere), linked to delirious beliefs about it [25]. The neglected body parts are ousted from mental representation of the body and the demonstrated existence of these parts is justified by confabulatory explanations [57, 58]. According to Vallar and Ronchi [59], the more frequent manifestations of somatoparaphrenia are: (1) the feeling of exclusion of the affected body parts, of their separation from the patient's body, (2) delusional beliefs of disownership of the affected body parts, (3) delusional beliefs about the affected body parts belonging to another individual, (4) complex delusional misidentifications of the affected body parts, and (5) the presence, in some cases, of associated disorders, such as supernumerary limbs, misoplegia, and personification.

At the brain level, somatoparaphrenia is frequently associated with right hemisphere damage. This observation suggests that the large base of the right hemisphere is involved in the sense of body identity and ownership. In particular, the involved neural circuitry consists of the temporo-parietal junction, posterior insula, basal ganglia, and insular cortex. In their review, Vallar and Ronchi [59] stated that a relevant mechanism of somatoparephrenia can be found within a network that includes the frontal premotor and posterior parietal cortices as well as subcortical structures such as the thalamus, basal ganglia, and superior colliculus [60]. This network is associated with defective multisensory integration rather than with the impairment of specific sensory modalities.

References

1. Synofzik M, Vosgerau G, Newen A (2008) I move, therefore I am: a new theoretical framework to investigate agency and ownership. Conscious Cogn 17:411-424
2. Marcel A (2003) The sense of agency: awareness and ownership of action. In: Roessler J, Eilan N (eds) Agency and self-awareness: issues in philosophy and psychology. Oxford University Press, Oxford, pp 48-93
3. Gallagher S (2000) Philosophical conceptions of the self: implications for cognitive science. Trends Cogn Sci 4:14-21
4. Gallagher S (2008) Direct perception in the intersubjective context. Conscious Cogn 17:535-543
5. Shoemaker S (1994) Self-knowledge and "inner sense". Philos Phenomen Res 54:249-314
6. Neisser U (1988) Five kinds of self-knowledge. Philos Psychol 1:35-59
7. Gazzaniga M (1995) Consciousness and the cerebral hemispheres. In: Gazzaniga MS (ed) The cognitive neurosciences. MIT Press, Cambridge, MA, pp 1391-1400
8. Sato A, Yasuda A (2005) Illusion of sense of self-agency: discrepancy between the predicted and actual sensory consequences of actions modulates the sense of self-agency, but not the sense of self-ownership. Cognition 94:241-255
9. Tsakiris M, Schütz-Bosbach S, Gallagher S (2007) On agency and body-ownership: phenomenological and neurocognitive reflections. Conscious Cogn 16:645-660
10. Gallagher S (2007) The natural philosophy of agency. Philosophy Compass 2:347-357
11. Legrand D (2007) Subjectivity and the body: introducing basic forms of self-consciousness. Conscious Cogn 16:577-582
12. Balconi M (in press) The sense of agency in psychology and neuropsychology. In: Balconi M (ed) Neuropsychology of the sense of agency. Nova Science, New York
13. Tsakiris M, Prabhu G, Haggard P (2006) Having a body versus moving your body: how agency structures body-ownership. Conscious Cogn 15:423-432
14. Botvinick M (2004) Probing the neural basis of body ownership. Neuroscience 305:782-783
15. Tsakiris M, Haggard P (2005b) The rubber hand illusion revisited: visuotactile integration and self-attribution. J Exp Psychol Human 31:80-91
16. Damasio AR (1994) L'errore di Cartesio. Emozione, ragione e cervello umano. Adelphi, Milano, pp 187-234
17. Serino A, Farnè A, Làdavas E (2006) Visual peripersonal space. In: Vecchi T, Bottini G (eds) Imagery and spatial cognition: methods, models, and cognitive assessment. John Benjamins, Amsterdam, pp 323-362
18. Schwabe L, Blanke O (2007) Cognitive neuroscience of ownership and agency. Conscious Cogn 16:661-666
19. De Vignemont F (2007) Habeas corpus: the sense of ownership of one's own body. Mind Lang 22:427-449
20. Bermúdez JL, Marcel A, Eilan N (eds) (1995) The body and the self. MIT Press, Cambridge, MA
21. Jeannerod M (2003) The mechanism of self-recognition in humans. Behav Brain Res 142:1-15
22. Lenggenhager B, Tadi T, Metzinger T, Blanke O (2007) Video ergo sum: manipulating bodily self-consciousness. Science 317:1096-1099
23. Bermúdez JL (1998) The paradox of self-consciousness. MIT Press, Cambridge, MA
24. Martin MGF (1995) Bodily awareness: a sense of ownership. In: Bermúdez JL, Marcel A, Eilan N (eds) The body and the self. MIT Press, Cambridge, MA, pp 267-289
25. Maravita A (2007) I disturbi della rappresentazione corporea [Deficits in body representation]. In: Vallar G, Papagno C (eds) Manuale di neuropsicologia [Manual of neuropsychology]. Il Mulino, Bologna, pp 201-221

26. Carruthers G (2009) Is the body schema sufficient for the sense of embodiment? An alternative to de Vignemont's model. Philos Psychol 22:123-142
27. Tsakiris M, Fotopoulou A (2008) Is my body the sum of online and offline body-representations? Conscious Cogn 17:1317-1320
28. Carruthers G (2008) Types of body representation and the sense of embodiment. Conscious Cogn 17:1302-1316
29. Graziano M, Botvinik M (2001) How the brain represents the body: insights from neurophysiology and psychology. In: Prinz W, Hommel B (eds) Common mechanisms in perception and action, attention and performance. Oxford University Press, Oxford, pp 136-156
30. Farrer C, Frith CD (2002) Experiencing oneself vs. another person as being the cause of an action: the neural correlates of the experience of agency. Neuroimage 15:596-603
31. Farrer C, Franck N, Paillard J, Jeannerod M (2003) The role of proprioception in action recognition. Conscious Cogn 12:609-619
32. Craig AD (2002) How do you feel? Interoception: the sense of the physiological condition of the body. Neuroscience 3:655-666
33. Berlucchi G, Aglioti S (1997) The body in the brain: neural bases of corporal awareness. Trends Neurosci 20:560-564
34. Shimada S, Hiraki K, Oda I (2005) The parietal role in the sense of self-ownership with temporal discrepancy between visual and proprioceptive feedbacks. NeuroImage 24:1225-1232
35. Blakemore SJ, Frith C (2003) Self-awareness and action. Curr Opin Neurobiol 13:219-224
36. Craig AD (2009) How do you feel-now? The anterior insula and human awareness. Nat Rev Neurosci 10:59-70
37. Roper SN, Levesque MF, Sutherling WW, Engel J Jr (1993) Surgical treatment of partial epilepsy arising from the insular cortex. J Neurosurg 79:266-269
38. Penfield W (1955) The role of the temporal cortex in certain psychical phenomena. J Ment Sci 101:451-465
39. Ehrsson HH, Spence C, Passingham RE (2004) That's my Hand! Activity in premotor cortex reflects feeling of ownership of limb. Science 305:875-877
40. Ehrsson HH, Holmes NP, Passingham RE (2005) Touching a rubber hand: feeling of body ownership is associated with activity in multisensory brain areas. J Neurosci 25:10564-10573
41. Rizzolatti G, Sinigaglia C (2006) So quel che fai, il cervello che agisce e i neuroni specchio [I know what you are doing, the acting brain and the mirror neurons]. Raffaello Cortina Editore, Milano
42. Rizzolatti G, Fogassi L, Gallese V (2002) Motor and cognitive functions of the ventral premotor cortex. Curr Opin Neurobiol 12:149-154
43. Rizzolatti G, Camarda R, Fogassi L et al (1988) Functional organization of inferior area 6 in the macaque monkey. Exp Brain Res 71:491-507
44. Vallar G (2007) A hemispheric asymmetry in somatosensory processing. Behav Brain Sci 30:223-224
45. Longo MR, Azañón E, Haggard P (in press) More than skin deep: body representation beyond primary somatosensory cortex. Neuropsychologia
46. Lopez C, Halje O, Blanke O (2008) Body ownership and embodiment: vestibular and multisensory mechanisms. Clin Neurophysiol 38:149-161
47. Blanke O, Landis T, Spinelli L, Seeck M (2004) Out-of-body experience and autoscopy of neurological origin. Brain 127:243-258
48. Brugger P (1997) Illusory reduplication of one's own body: phenomenology and classification of autoscopic phenomena. Cogn Neuropsychiatry 2:19-38
49. Blanke O, Arzy S, Landis T (2008) Illusory perceptions of the human body and self. In: Goldberg G, Miller B (eds) Handbook of clinical neurology. Neuropsychology and behavioral neurology, vol. 88. Elsevier, Paris, pp 429-458

50. Maravita A (2006) From body in the brain, to body in space: sensory and motor aspects of body representation. In: Knoblich G, Shiffrar M, Grosjean M (eds) The human body: perception from the inside out. Oxford University Press, Oxford, pp 65-88
51. Halligan PW, Marshall JC, Wade DT (1993) Three arms: a case study of supernumerary phantom limb after right hemisphere stroke. Brit Med J 56:159-166
52. Botvinick M, Cohen J (1998) Rubber hand "feel" touch that eyes see. Nature 391:756
53. Denes G, Cappelletti JY, Zilli T et al (2000) A category-specific deficit of spatial representation: the case of autotopagnosia. Neuropsychologia 38:345-350
54. Buxbaum LJ, Coslett HB (2001) Specialized structural descriptions for human body parts: evidence from autotopagnosia. Cogn Neuropsychol 18:363-381
55. Schwoebel J, Coslett BH, Buxbaum JL (2001) Compensatory coding of body part location in autotopagnosia: evidence for extrinsic egocentric coding. Cogn Neuropsychol 18:363-381
56. Guariglia C, Antonucci G (1992) Personal and extrapersonal space: a case of neglect dissociation. Neuropsychologia 30:1001-1009
57. Ramachandran VS (1995) Anosognosia in parietal lobe syndrome. Conscious Cogn 4:22-51
58. Bisiach E, Geminiani G (1991) Anosognosia related to hemiplegia and hemianopia. In: Prigatano GP, Schacter DL (eds) Awareness of deficit after brain injury. Oxford University Press, Oxford, pp 17-39
59. Vallar G, Ronchi R (2009) Somatoparaphrenia: a body delusion. A review of the neuropsychological literature. Exp Brain Res 192:533-551
60. Vallar G, Maravita A (in press) Personal and extra-personal spatial perception. In: Berntson GG, Cacioppo JT (eds) Handbook of neuroscience for the behavioral sciences. Wiley, Hoboken, New Jersey

Subject Index

A
Action
 - identification theory 136, 139, 161
Anarchic hand syndrome 10
Anosognosia 75, 136, 130, 131, 180
Autism 125, 135
Autobiographical memory 64
Awareness
 - of action 3-6, 11, 12, 51-53, 69, 98, 135
 - of intention 11, 13, 15, 51, 52

B
Bandura 82, 83, 89, 99, 108
BIS/BAS 138
Blindsight 52, 125, 128
Body
 - ownership 51, 75, 126, 177, 173-175, 177, 178, 180-186
 - representation 57, 126, 128-131, 178-180, 182, 183, 185
 - schema 119, 129, 131, 176, 177, 179, 185, 186
Bratman 38-40, 110, 111, 113, 114

C
Cerebellum 4, 15, 56, 57, 70, 71, 73, 74, 82, 117, 118, 128, 135, 181, 182
Collective intentions 109, 110, 120
Comparator model 7, 8, 56, 72, 73, 117, 119, 132, 134, 145-148, 150, 154
Consciousness 3, 11, 12, 14, 16, 18, 20, 48, 50, 52, 54, 60-62, 75, 81, 85-88, 98-100, 107-109, 126, 173-175, 177, 178, 181
Control
 - strategies 7
Coopertative-activity 113

D
Dennett 5, 63, 174
Detached awareness 12, 50

Disownership phenomena 184
Disruption 7, 59, 62, 86, 98, 99, 123, 125, 127, 134-139, 162, 177, 182, 186
Dorsal visual pathway 54

E
Empathy 18, 19
ERN 138
ERP 15, 18, 52, 137, 138
Execution, intentions 100

F
Feeling
 - of a presence 183,
 - of agency 47, 49, 50, 51, 57, 60, 135, 148-150, 177
fMRI 52, 55, 68, 70, 71, 76, 117, 118, 181
Forward model 7, 13, 56, 62
Free choice 12, 59
FRN 138-140
Frontotemporal dementia 132

G
Gallagher 19, 26, 27, 64, 108, 116, 173-175

H
Haggard 9, 10, 51

I
Illusion
 - of agency 11, 12
 - of control 5, 164-166
 - of intention 58
Immersed awareness 50
Immunity principle 61, 62, 174
Initiation 3, 6, 9, 15, 20, 75, 81, 87
Insula 4, 10, 59, 60, 70, 71, 74, 75, 82, 118, 180-182, 186

Intentional binding 6, 9, 10, 81, 151
Intentions 6, 7, 9-12, 15, 19, 24-28, 33-35, 37-40, 47, 48, 51, 52, 54, 57, 58, 62, 75, 76, 81, 82, 87, 98, 99, 109-113, 115-117, 119, 125, 127, 132, 134, 137, 158, 160, 177
Interaction 4, 6, 24, 26, 27, 48, 50, 54, 65, 76, 83, 88, 91, 94-99, 107, 109-113, 115, 116, 118, 119, 130, 132, 178, 184, 185
Inter-agency 107
Intersubjectivity 111, 115

J
Joint-action 107, 112, 113, 119
Joint-agency 114, 119
Judgment of agency 47, 50, 51, 135, 175

L
Libet 15, 16, 51, 58

M
Minimal self 61, 62, 108, 126, 173
Mirror neuron system 8, 113
Moral responsibility 48
Movement 4-6, 9,14-16, 18-20, 23, 29, 31, 33-35, 37-39, 51-57, 59, 60, 62, 70, 71, 74, 75, 81, 82, 112, 117, 127, 132, 133, 136, 147, 152, 161, 162, 164, 175, 179, 181, 184
Multimodal representation 178, 179

N
Narrative self 14, 47, 61, 63, 64, 148, 173, 174
Neglect 125, 129-131, 185, 186
Neurological disorders 182
Numbsense 128, 179

O
Obsessive-compulsive disturb 125
Ownership 6, 11, 19, 51, 57, 60, 62, 63, 75, 108, 125-127, 129, 131, 132, 153, 173-186

P
Parietal cortex 4, 15, 56, 57, 59, 70-74, 82, 117-119, 128, 130, 181
Perception 6, 8, 11, 18-20, 23-41, 52, 55, 57, 60-62, 74-76, 83, 87, 94, 98, 99, 111-113, 116, 119, 125, 126, 129, 130, 134, 138, 146, 150, 152, 161, 164, 166, 167, 177-180, 182-185
Perseveration 131
PET 52, 69-71, 117
Posterior parietal lobe 181

Prefrontal cortex 4, 15, 48, 49, 57, 60, 70, 71, 75, 82, 160
Priority principle 12, 16
Proprioceptive 6, 11, 53, 54, 57, 60, 74, 87, 117, 126-1280 167, 175, 177, 181, 182, 184, 185

R
Readiness potential 15, 16, 51, 52
Rubber hand illusion (RHI) 60, 127, 177, 179, 181, 182, 184

S
Schizophrenia 5, 62, 76, 87, 117, 118, 125, 132-135, 145-148, 150-153
Self ascription 63, 126, 131, 175
Self-knowledge 63, 125, 126, 137, 175
Sense
- integration 57
- of agency 3-12, 14, 15, 18, 23, 24, 47-53, 59-64, 69, 70, 72-77, 84-84, 88-91, 96, 98-100, 108, 109, 111, 114, 116, 117, 120, 125, 126, 129, 131-135, 137-139, 145, 147-152, 154, 157, 158, 160, 161, 163, 165, 173-175, 181, 184
- of body ownership 75, 126, 174, 175, 177, 181, 182, 184, 185
Sensorimotor processes 47, 48, 175, 176
Shared collective-activity 107
Social cognition 19, 36, 113, 119, 135
Somatoparaphrenia 130, 185, 186
Somatosensory
- cortex 55, 73, 181
- neglect 129, 130
Spatial content 178, 179
Supplementary motor area 4, 52, 71, 75, 82, 118, 127

T
Theory of mind 24, 27, 40, 135
Thought insertion 62, 131, 133
TMS 51, 69, 72, 117

V
Ventral pathway 54
Visual
- feedback 6, 53, 57-59, 72, 74, 75, 134, 148, 152, 153, 176, 182
- neglect 129

W
Wegner 10, 12, 13, 16, 58, 161, 162